SpringerBriefs in Electrical and Computer Engineering

More information about this series at http://www.springer.com/series/10059

Shalli Rani · Syed Hassan Ahmed

Multi-hop Routing in Wireless Sensor Networks

An Overview, Taxonomy, and Research Challenges

 Springer

Shalli Rani
Department of Computer Science
SSD Women's Institute of Technology
Bathinda, Punjab
India

Syed Hassan Ahmed
School of Computer Science and
  Engineering
Kyungpook National University
Daegu
Republic of Korea

ISSN 2191-8112   ISSN 2191-8120 (electronic)
SpringerBriefs in Electrical and Computer Engineering
ISBN 978-981-287-729-1   ISBN 978-981-287-730-7 (eBook)
DOI 10.1007/978-981-287-730-7

Library of Congress Control Number: 2015948727

Springer Singapore Heidelberg New York Dordrecht London

Printed on acid-free paper

Springer Science+Business Media Singapore Pte Ltd. is part of Springer Science+Business Media
(www.springer.com)

Preface

Over the last decade, Wireless Sensor Network (WSN) has become very popular research field due to development of low-cost sensors. This network can operate even in harsh environments. Sensors when once deployed cannot be replaced and recharged, due to which WSN is currently facing many challenges. Today, users are interested in timely availability and reliability of data. To provide an insight into the solutions, a thought can be given via hardware and software. Nevertheless, hardware solutions can be costly in setup and difficult to maintain. To cope up with the current requirements, many other solutions are in trend such as development of efficient routing algorithms, collision-free routing protocols, congestion control algorithms, chain-based routing protocols, and energy-efficient routing protocols. However, none of these protocols provides optimum solution for the current challenges. All protocols have their own pros and cons in terms of complexity, scalability, energy constraints, delay tolerance, etc. WSN must employ distributed algorithms to support all the applications due to short-range communication and high-energy consumption features. It must be self-configurable, scalable, and robust. It has provided quite wide application portfolio for different branches such as military, transport, agriculture, industry, and health care. In future, stronger WSN application assortment is expected. In order to make this expansion possible, it is necessary to continually work on the solving of typical questions/problems related to the WSN development, e.g., standardization of communication protocols, the lack of energy-efficient power sources, and the development of new ultra-low-power microelectronic components.

The problematic of WSN is one of actual activities getting to the fore in the European Research Area since the issue of sensor networks was covered through "IoT" in FP7 program and strong continual extension is planned to be included also in Horizon 2020 program, especially in sections such as Smart Transport, Health, and Climate Action covered under Societal Challenges Pillar. Researchers to overcome the problem of energy scarcity and short-range communication of sensors have put enormous efforts forth. It has been realized that routing protocols, which provide multi-hop communication, are more robust and they are showing good results in terms of energy, scalability, and network lifetime. Multi-hop communication is not a new concept having been around for over twenty years, mainly exploited to design tactical networks. The simplest communication is a peer-to-peer communication formed by a set of stations within the range of each other that dynamically configure themselves to set up a temporary single-hop network. Bluetooth piconet is the most widespread example of single-hop networks. 802.11 WLANs can also be implemented according to this paradigm, thus enabling laptops' communications without the need of an access point. Single-hop networks just interconnect devices that are within the same transmission range. This limitation can be overcome by exploiting the multi-hop paradigm. In this networking paradigm, the sensors form network and cooperatively provide the functionalities that are usually provided by the network infrastructure. Nearby nodes can communicate directly by exploiting a single-hop wireless technology (e.g., Bluetooth, 802.11), while devices that are not directly connected communicate by forwarding their traffic via a sequence of intermediate devices. To turn on more advantages of WSN, we should move to a more pragmatic scenario in which multi-hop communication is used as a flexible and "low-cost" extension of wireless communication.

Therefore, authors would like to provide the overview of recent development in multi-hop routing protocols. It will follow an introduction to the various classifications of routing protocols. Pros and cons of each category will be enlisted. Current research on the various categories of multi-hop routing protocols is given to help the researchers to fine out classification for their protocols. Beginners can make themselves aware about the current trends by the overview of classifications of routing techniques. It is beneficial for the students who being involved in technical studies. The aim of this book is to present some of the most relevant results achieved by applying an algorithmic approach to the research on multi-hop routing protocols. The unique aspect of the book is to present measurements, experiences, and lessons obtained by implementing multi-hop communication prototypes.

Acknowledgements

My thanks go to everyone who supported me time to time in my research work. Firstly I would like to thank Dr. Jyoteesh Malhotra, GNDU Regional Campus, Jalandhar, Punjab, India and Dr. Rajneesh Talwar, CGC technical Campus, Jhenjeri, Punjab, India. Without their valuable guidance and support, this achievement would not be possible. They make me understand the basic problems they faced in the research and enable me to reach the solution.

Syed Hassan Ahmed from KNU, Republic of Korea is my co-editor for this book and his dedication towards the research and serving research community inspired me to pursue this book. It was nearly impossible to make this book published without his valuable efforts and time. I wish him best of luck for his future.

Last but not least, my gratitude is due to my mom, my sister, my son and my beloved one for their trust in me and for making me confident that I can put efforts to get success in each and every field.

Shalli Rani

First and foremost, I would like to thank my wife Aiemun Hassan for standing beside me throughout my research career so far and writing this brief book. She has been my inspiration and motivation for continuing to improve my knowledge and move forward in my career. She is my lifetime achievement, and I dedicate this first book of mine to her.

Moreover, without the prayers and best wishes of my late mother "Riaz Begum", all of my achievements so far and forever would never be possible. Here, I am also thankful to my co-author Shalli Rani, who really helped me in every aspect in the preparation of this book. Her prompt response and care to the literature and contribution is remarkable and sensational. She is a great and responsible researcher indeed.

I would also like to thank the Springer staff and editors, who actually helped a lot, and without their quick and efficient efforts, it was not possible to get our book published.

Syed Hassan Ahmed

Contents

1 Introduction 1
  1.1 Wireless Sensor Networks: An Overview 1
      1.1.1 WSN Challenges 2
      1.1.2 Applications of WSN 3
      1.1.3 Objectives and Design Issues of WSN 5
  1.2 Protocol Stack and Architecture 7
  1.3 Classification of Routing Protocols 10
  1.4 Limitations and Advantages of Various Schemes 11
  1.5 Future Trends in Routing: Multi-hop Routing Categories 12
  References 12

2 Multi-hop Energy Efficient Routing 15
  2.1 Introduction 15
  2.2 Multi-hop Energy Efficient Routing Protocols 17
      2.2.1 Chain Based Data Transmission 18
      2.2.2 Heterogeneity-Based Protocols 21
  2.3 Comparative Analysis 24
  2.4 Summary and Future Trends 27
  References 27

3 Multi-hop Reliability and Network Operation Routing 29
  3.1 Brief History 29
  3.2 Reliability and Network Operation Based Protocols 31
      3.2.1 Multipath-Based Protocols 32
      3.2.2 Query Based Protocols 33
      3.2.3 QoS-Based Routing 35
      3.2.4 Negotiation Based Protocols 37
      3.2.5 Coherent Based Protocols 39
  3.3 Comparison 40
  3.4 Summary and Future Trends 43
  References 43

4   Multi-hop Network Structure Routing Protocols   45
    4.1   Introduction   45
    4.2   Network Structure Based Routing Protocols   46
          4.2.1   Flat Structure Routing Protocols   47
          4.2.2   Location Based Protocols   49
          4.2.3   Hierarchical Based Protocol   51
    4.3   Comparison   54
    4.4   Summary and Future Trends   57
    References   57

5   Future Research and Scope   59
    5.1   Fundamentals of WSN   59
          5.1.1   Comparative Analysis of Multi-hop Routing Protocols   61
    5.2   Future Scope   64
    5.3   Conclusion   65
    References   66

Index   69

Chapter 1
Introduction

Abstract Advancements in wireless sensor networks have led the researchers to develop new protocols to cater all the stringent requirements of WSN. Awareness in routing protocols has been considered a most important issue because they are dependent upon the applications design and architecture. WSN systems gather data from multiple sensors to monitor some area. Networking and management of these sensors is difficult task due to various constraints imposed on tiny nodes. Routing is very challenging job in WSN due to contemporary communication. In this chapter, we have thrown light on the various challenges and design issues in WSN. A general classification of routing protocols is presented here with advantages and limitation of each scheme. Overview of applications of WSN and future trends towards the multi-hop routing protocols is discussed here with various research issues.

Keywords WSN · Design issues · Challenges · Applications · Architecture · Routing

1.1 Wireless Sensor Networks: An Overview

With the advancement in development of smart and micro sensors, Wireless sensor networks (WSN) have attained worldwide attention and have been considered as the most important technology of this century. WSN is also known as wireless sensor and actor network (WSAN) [1]. It is consisted of autonomous sensors, which are spatially dispersed to monitor environmental or physical conditions such as sound, pressure, temperature etc. These tiny and cheap sensors are equipped with small battery, low memory and have limited processing capability. Sensors collaboratively perform the duty of sensing and gathering information from the environment and transmitting it to the main station. Need of WSN came into existence with military applications, and now it is used in many fields like bordering surveillance, health care monitoring, Industry process control, watering

© The Author(s) 2016
S. Rani and S.H. Ahmed, *Multi-hop Routing in Wireless Sensor Networks*,
SpringerBriefs in Electrical and Computer Engineering,
DOI 10.1007/978-981-287-730-7_1

monitoring etc. Applications of Wireless sensor networks require the techniques of wireless ad hoc networking. Many protocols and algorithms have been proposed for the ad hoc network (MANET) but they are not suitable for WSN. A few elementary differences between Wireless Sensors Networks and Mobile Ad hoc Network (MANET) can be outlined on the basis of following characteristics [2]:

1. Global Identification: Identification at the global level is not preferred in WSN (Increases overhead at run time) due to the vast number of applications and nodes but it is required in MANET.
2. Data centric: Redundancy is required in WSN but this concept is avoided in MANET until sharing of the some file is required.
3. Scalability: Most of the applications require billions of the nodes to be deployed, making the network denser which makes it dissimilar from MANET.
4. Soundness and Quality of Service metrics (QoS): MANET is more reliable than WSN and reliability per node is necessary at fair level but in WSN requirements of QoS metrics is different because applications using WSN must be more energy efficient as batteries of sensor nodes cannot be replaced once they are deployed.
5. Fault-tolerant: WSN is expected to work even after the failure of large number of nodes, which is result of restricted battery capacity of sensors. So more attention is needed in case of WSN to make it fault tolerant as compared to traditional network.
6. Network size: In different types of applications, network size will vary in WSN according to their requirements, which lead to the development of different protocols for dissimilar applications, but this is least necessary in MANET.
7. Operating Software: WSN's have limited memory and processing capabilities so operating software must be simple but in MANET complexity, software (heavy weight routing protocols) can be used.
8. Traffic patterns: MANETs have more conventional traffic patterns. In contrast to that, WSNs tend to have low data rates for long periods intervened by bursts of data flows and high data rates are frequents in case of some events.

Due to the above differences between MANET and WSN, existing protocols and algorithms of MANET cannot be applied to WSN. Many research tricks have been carried out all over the world to solve various application and design issues of WSN and significant advancements are perceived in its development and deployment. In the upcoming days, WSN will bring revolution in our interaction with the world, in our day-to-day activities, in our ways of living etc. [3].

1.1.1 WSN Challenges

Many research challenges can be addressed to produce realistic WSN such as: Robust system operation, Security, from raw data to knowledge, Privacy, Openness and heterogeneity, Privacy, Real-time control and operation etc. [4]. Each challenge has to deal with strident, hesitant and growing environments. The unique features of WSN pose many design challenges and can be summarized as below:

1. Ad hoc and Random Deployment of Nodes: Many WSN's are comprised of few thousand nodes to few lakh nodes or several more. Deployment of the nodes can be static (manual) or random. These nodes are deployed randomly over the antagonistic region where they have to organize themselves automatically for communication. They need to set up network before exchange of any information.
2. Unattended and dynamic environment: Sensor nodes are deployed randomly in the harsh and dynamic environment. In this environment some sensor nodes may cause failure of network due to early depletion of energy or due to some other environment conditions. Fault tolerance should be considered significant while developing the protocols for WSN applications.
3. Limited capacity and processing capability: Sensor nodes are battery operated and they have very limited battery capacity. This feature has presented many new challenges in advancement of software and hardware. To deal with limited energy capacity of sensor nodes new research is required not only in the design of communication protocols and network but also in terms of hardware.
4. Various requirements of applications: There are numerous WSN applications developing each day and all have different requirements. Due to diversity in their requirements, different design of the network, different communication protocols and different strategies are required to be developed.
5. Unreliable Communication: Wireless links are affected by the interference signals (e.g. SNR). Due to these signals, communication among the wireless sensor nodes becomes unreliable. And it can be broken at anytime. It also suffers from the range problem. This presents the new challenge in front of WSN, which is required to be dealt with some finer solutions.

1.1.2 Applications of WSN

WSN can be applied to any type of environment, since it reduces the delay and cost in deployment as compared to traditional sensor networks which are impossible to deploy e.g. in deep oceans, combat zones, outer space, unreceptive environment. Nevertheless, the development and availability of low cost sensors has assured the wide range of applications in upcoming days. In this section we have discussed some of the WSN applications.

1. Environmental Applications: Environment Monitoring can be classified into two groups (a) indoor monitoring (b) outdoor monitoring [5]. It is expected that this application will require deployment of WSN as the substantial identifiers that must be monitored, are e.g., temperature, air and humidity. It covers the large area and distributed over the entire region.

(a) Indoor Monitoring: Indoor monitoring applications are naturally comprised of buildings and offices monitoring. These applications are engaged in sensing air quality, humidity, light, and temperature, and other important indoor applications may include fire and civil structures deformations detection.

(b) Outdoor Monitoring: These applications include weather forecasting, volcano eruption, chemical perilous exposure, earthquake detection, habitat supervision, traffic monitoring, and flooding detection etc. Sensor nodes are found to be very advantageous in the field of agriculture. In agriculture, temperature and soil moisture monitoring is the most important application of WSN.

One of the most motivating and beneficial applications of WSNs is the ability to look at the giant picture of the monitor environment. There have been many implementations of macroscopes for the same purpose, for example, a WSN deployed for a wildlife monitoring site on Great Duck Island, on Redwood trees, tracking zebras in their natural habitat, Luster: an environmental science application for measuring the effect of sunlight on under shrub growth on barrier islands etc. [4].

2. Health Care Applications: Sensor have been used in medicines and public health from a long time [6]. Some of the healthcare applications are as follow [7]:

(a) Vital sign supervision in hospitals: Sensors are being used to monitor the patient while roaming in the hospital, which was not possible with the wired sensors.

(b) Monitoring in mass—casualty catastrophe: Due to the portable and scalable nature of wireless sensing systems, they can be used to report the levels of victims and to track the health status of responders at the disaster site.

(c) At-home and portable aging: WSN, deployed in people's living places or in person can gather information about behavioral and physiological states in real time and all over the place. This can be useful for the self-awareness and analysis. Early detection and early intervention is possible through this system.

(d) Support with sensory and motor decline: System of sensors can assist and guide the patients with declining the sensory and motor capabilities. Wireless networked sensing facilitate novel category of assistive devices such as walking navigation and way finding for the persons who are visually impaired.

(e) Behavioral and Medical studies: Sensors situated in body together with sensors equipped with smart phones have revolutionized the research in behavioral and physiological sciences.

3. Industrial Applications: WSN plays a significant role in creating self-remedial and reliable system that actively and quickly responds to real time events with corrective steps [8]. Maintenance of the machinery in industries can be arranged

via WSN. Sensor nodes have been used in many industry applications like coal mine, oil industry, gold and diamond scrutiny etc. Sensors can monitor condition of the pipelines in coal refinery and chemical plants. Lives in industries can be made safer by help of sensors as they can quickly respond before any machine failure.

4. Area Monitoring Applications: The additional trait of Wireless integrated networked sensors (WINS) robustness and self-organization makes it deployable by inexpert troops in fundamentally several situations. Rockwell is mounting a chain of many consumer, military, industrial and aerospace applications for WINS [9]. WINS will dramatically improve troop safety for the urban terrain, as they monitor and clear buildings, rooftops and intersections by providing incessant caution. Example of area monitoring in the military application is to detect the intrusion by enemy and in civil application is geo fencing of oil or gas pipelines [10]. When location aware device enters and exits a location, the device is notified with message. Global positioning system is another application of area monitoring. GPS project was began in 1973 to overcome the drawbacks of previous navigation systems, which included the ideas from many predecessors and engineering design studies from 1960s [11]. The system developed by department of defense (DoD) used the 24 satellites and became operational in 1995.

5. Entertainment: WSN has been used in music industry to sense the live music performers e.g. audio cubes. Audio cubes are collection of intelligent wireless emitting objects, which can detect each other's location and gestures of users [12]. Bert Schiettecatte created it. It is example of ambient device and concrete user interface.

6. Military Applications: WSN came into existence with the requirements of military application such as object protection, battlefield monitoring, remote sensing, and intrusion detection. Sensors are deployed randomly to detect the weapons, to monitor the presence of vehicles, to track the movements of opponents etc. In future, WSN will play an important role for FAAD C31 military system.

1.1.3   Objectives and Design Issues of WSN

Various methods of routings are proposed from the past to the present can be classified based on the objective of the routing, nature of the applications, model of the operation and on the network architecture. Purpose of routing algorithms in the literature varies in the literature. Mostly objectives are set to meet the requirements of the applications. For example if the application is susceptible to the time (Real time applications) then time delivery, shortest paths for routing are considered for the data transmission and connectivity of the nodes is required for reliability. Here we have discussed various objectives and design issues of wireless sensor networks.

(a) Network lifetime maximization: Network lifetime is the major constraint of WSN, which is based on the battery, operated low power sensor nodes. For the environmental applications; where it is not possible to replace the sensor nodes this issue is of main concern. It is required that in intra-cluster communication (communication between cluster heads and Nodes) energy consumption should be minimized [13]. Distance between cluster heads (CHs) and nodes should be minimized to gain the benefits of the energy conservation [14, 15]. Adaptive networking can be used for the longevity of the network lifetime [16, 17]. By minimizing the load, route set up and by inter and intra communication among the nodes and the CHs; lifetime of the network can be maximized [18].

(b) Fault tolerant: In the environmental applications of WSN due to harsh conditions some nodes can discontinue operation due to malfunctioning and physical damage which lead to the loss of important data of the cluster heads. Most efficient way to recover from this damage is re-clustering of the network [19]. But this process can expand a lot of energy so re-election of CHs or the back-ups of the CHs is the most prominent technique in the literature. Fault tolerance can also be achieved by rotating the role of CH among nodes and this can further help in load balancing [20].

(c) Timely delivery: Existence of routing path among the nodes, Cluster heads (CHs) and Base station (BS) is necessary for the successful transmission of data. Unless the cluster heads are more powerful for long-range transmission like satellite link; inter cluster communication is required in many applications [21]. When delay tolerance is the design issue then intra-clustering communication is also required which is factored by the setting up the 'H' number of hops permitted on the routing path. In 'H' number of hops problem 'H'–hop clustering is proposed to solve the time delivery issue [22–24].

(d) Load Balancing: To meet the aim of the best performance it is perceptive to balance the load among the duties assigned to the sensor nodes [25]. Even distribution of load; among the nodes is required for significant communication at the intra level. To extend the network lifetime and to avoid the hot spot problem (where some nodes cannot communicate with BS due to exhaustion of energy by the CHs); it become crucial to set up the equal size clusters. Duty of CHs is to aggregate data so it is essential to have same number of nodes in all the clusters so combined data could be organized at the same time for further processing without making any delay.

(e) Reduction in the redundancy: In the applications where the sensor nodes are randomly deployed some nodes may generate the redundant data, data aggregated nodes will have multiple copies of the same data, which can put unnecessary load on the network. Data can be aggregated by using functions like repression (to remove redundant data), average, min [26]. These functions can allow the sensor nodes to perform in-house reduction of data. Reducing the duplication locally before transmitting to the CHs saves a lot of energy than at the data aggregation step. This method can also avoid the extra traffic on the network thus remunerations in terms of congestion control and the energy savings; can be achieved. Role of the CHs can be rotated among the nodes to balance the load or for better clustering scheme to exploit the benefits of extra processing [27]. It has been found in the literature that data aggregation is possible through signal processing capabilities, known as data fusion where accurate signals can be produced by beamforming or by reducing the noise. The load should not overburden cHs as it is intuitive to expect that CHs will perform the aggregation or fusion of data so number of sensor nodes should be limited.

(f) Minimization of the number of Clusters: In the heterogeneous cluster based networks where some nodes are more powerful than others; a network designer would like to employ minimum number of nodes as they are more expensive than normal nodes. In the applications like border security surveillance, military scouting, it is undesirable for the nodes to be noticeable so deployments of more nodes as CHs whose size is larger than other nodes is avoided. If there are more number of clusters then more number of CHs will be required to gather data which will utilize more energy due to extra processing in gathering data and other functions. To conserve energy optimal number of cluster heads should be elected [28].

(g) Data Communication and data Collection via Single hop or Multi-hop: Sensor nodes transmit their data to the sink or the Base station by direct (Single Hop) or indirect (Multiple Hop) transmission. Data can be transmitted in the continuous manner or occasionally depending upon the requirement of the application. Data can be sent in broadcast, point to point or the local fashion. In point to point an arbitrary node can forward data only to a single node in contrast to broadcast scheme in which a single node can transmit data to many nodes at a time. In the local communication data from data is exchanged between two nodes to get the status of each other. Multiple nodes transmit data to the single node is known as convergence which is used for data collection and commands from the BS are routed to all the nodes for divergence. Data can be processed before or after the aggregation. Innovative ideas need to be generated to compensate among different techniques to maximize the benefits of novel routing algorithms.

1.2 Protocol Stack and Architecture

The sensor nodes are deployed randomly in a field and they have capabilities to gather data and to transmit it to the sink. Data is routed to the sink via single hop or multi hop communication. The sink communicates with sensor nodes with the help of Internet or satellite [29]. The protocol stack used for sink and all the senor nodes can be observed from Fig. 1.1. Along with five popular layers, there are three other layers: mobility management plane, power management plane and task management plane. These layers provide additional functionality to the sensor nodes. These layers help the nodes to reduce the overall energy consumption and consequently

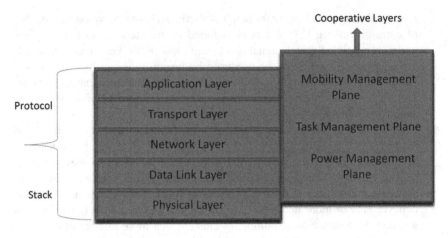

Fig. 1.1 Protocol stack for WSN

increase the network lifetime. The power management takes care of the duties related to power. For example, if power of any node falls below the minimum threshold level, then it can broadcast about its power and can deny participating in transmitting the data. So that remaining power can be used for sensing the data. The mobility pane holds the responsibility of the mobile status of the nodes so that it could be known which nodes are the neighbor nodes to cooperate in transmission of data. The task pane helps in scheduling the tasks of the nodes. As nodes falling in the same region can turn off their radios except few to sense the data at particular time. This duty can be exchanged alternatively among the nodes for energy conservation. The task pane also controls sharing of channel among the nodes. From WSN point of view, if sensor nodes will work collaboratively then it would be more efficient in terms of network lifetime. The WSN trails the architecture as shown in Fig. 1.2. Sensor nodes from s1 to s4 cooperate with each other to transmit the data to the sink.

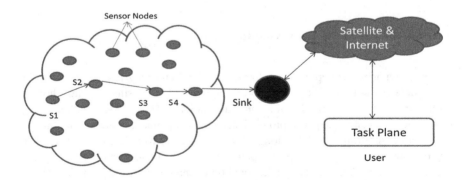

Fig. 1.2 Communication of sensor nodes with sink

The protocol stack of WSN is comprised of five layers as shown in Fig. 1.1: Physical layer, Data Link layer, Network layer, Transport layer and Application layer with three cooperative layers. The application layer deals with various application protocols. Transport layer takes care of reliable data delivery. Network layer handles the routing tasks to route the data to the transport layer. Data link layer is accountable for data multiplexing, data reception, error handling, and channel access. Physical layer is responsible for signal modulation and demodulation, signal transmission and reception, signal generation etc.

1. Application Layer: Time synchronization, node localization, network security etc. functions are provided by various protocols as application layer. Sensor management protocol (SMP) offers various software related services such as querying the status of sensor nodes, moving sensor nodes, exchanging the data among the nodes etc. The sensor query and data dissemination protocol (SQDDP) presents response to queries and issues queries to other nodes [29]. The sensor query and tasking language (SQTL) awards the language for sensor programming.

2. Transport Layer: Although this layer is responsible for end to end data delivery but traditional transport layer protocols responsible for end to end data delivery are not suitable for WSN. Window based Congestion control algorithms and error control algorithms cannot be applied directly to the WSN due to their low memory, limited energy and capacity. All applications of sensor network have their own design issues and requirements. And they need new protocols to be developed for them according to their needs.

3. Network Layer: This layer provides all the service required for routing of data from source node to the sink. The data may be transmitted via single hop or multi-hop communication. But main objective of this layer is conserve energy while transmitting the data. Many protocols are developed and proposed for this layer, and major aim of those protocols is to route the data in energy efficient manner without causing much delay.

4. Data Link Layer: Major design purpose of data link layer is management of medium access control (MAC). Its duty is to divide the channel access among the sensor nodes fairly for the optimum network throughput and timely delivery of data. Another services provided by data link layer are error control, encoding and decoding. Due to these services data link layer protocols have to become favorable among less complexity, energy efficiency or less processing.

5. Physical Link layer: This layer accepts and transmits the signals and converts them to the bit stream or vice versa. Physical layer provides signal modulation, demodulation, frequency generation, encoding and decoding services. This layer also requires management with underlying hardware. Selection of frequency and medium for communication amid the nodes is another common problem for physical layer.

1.3 Classification of Routing Protocols

In general routing protocols can be divided into—flat based routing, hierarchical-based routing, and location-based routing under the class of network operation, multipath-based, query-based, negotiation-based, QoS-based, or coherent-based routing techniques reliant on the protocol operation [30], multi-hop, heterogeneity based and chain based under the category of energy efficiency [31] as shown in Fig. 1.3.

All sensor nodes are allocated equal roles or functionality in flat based routing. Nevertheless, in hierarchical routing all nodes play different roles in the network. Locations of the nodes are exploited in location aware routing. In adaptive routing, certain parameters are reserved in control to handle the different network conditions and for energy efficiency. The route establishment from the node to the sink can be classified as reactive, proactive and hybrid routing. In reactive routing, the route is established on claim that is routes are dynamic. In proactive routing, the routes are established and decided before the transmission of data (routes are static). In dynamic selection of routes, the more energy of nodes is consumed in searching of routing paths. So some protocols use hybrid routing, in which static paths are elected before the transmission of data but new routes can be launched according to the requirements, and need of nodes. In cooperative or chain based routing, nodes

Fig. 1.3 General classification of routing protocols

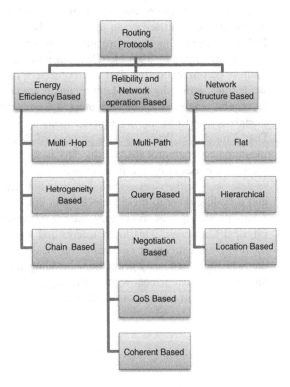

help each other in transmitting the data. In this book, we have classified the routing protocols on the basis of multi-hop transmission, which is further elaborated in subsequent chapters.

1.4   Limitations and Advantages of Various Schemes

All categories explained above in Sect. 1.3, have their own drawback and advantages. In energy efficient protocols, routing is destination oriented and directional. The nodes in the transmission path collect energy metrics. The routing path, which consumes less energy, is selected for the transmission of data and this choice of path is computed at every node and for every packet. Energy aware routing has many advantages as traffic is spread over the different paths to avoid congestion and retransmission. Redundancy is controlled while keeping the maintenance of the energy efficient paths. Hot spot problem, which arises due to early depletion of the energy of the nodes near the sink, is mitigated. Various paths are checked continuously for energy efficient routing. But, searching of paths again and again adds more delay in transmission and hence new trade off is raised between the energy and the delay.

In location aware routing, the position of nodes is known. Nodes can transmit data easily to their neighbor nodes. Data is transmitted easily in cooperative way. Services of location are routing more beneficial for mobile nodes. As there position can be easily traced and followed. Information about the new neighbor nodes can easily be explored and used. Nodes positions are exploited for routing. Distance estimation is possible for the incoming data. But to access the location of nodes require received signal strength (RSS) to be examined or to use the global positioning system (GPS) antennas. GPS is very costly techniques to be employed for all the sensor nodes and examination of RSS is only estimation. It can lead to deplete energy up to large extent.

In the network operation based protocols, data is transmitted by negotiation or routing is based on the query. However, routing can be only source initiated or destination oriented. QoS metrics are considered in this type of routing but trade-offs among all the metrics have to be examined carefully to exploit all the benefits of WSN.

Routing on the basis of network structure is beneficial in terms of network throughput, delay and energy efficiency. But in flat routing protocols, network throughput is increased on the cost of more energy expenditure. In hierarchical networking, delay is introduced while conserving energy for the network lifetime.

Routing protocols require careful examination to fit the advantages to WSN applications according to their needs.

1.5 Future Trends in Routing: Multi-hop Routing Categories

Data to the sink can be transmitted via single hop or multi hop communication. All the sensor nodes can use single hop communication but in long distance transmission, the energy consumption is much higher in transmission as compare to processing and sensing tasks. Transmission energy dominates the overall energy used in communication process. The requirement of energy goes on increasing with the increase of distance [32–34]. Therefore, it becomes necessary to reduce the energy consumption and to enhance the network lifetime. Therefore, it is preferable to use short-range multi hop communication. In multi hop communication, all nodes communicate with each other using wireless channels without need of any control structure and common infrastructure. Nodes cooperate with each other to forward the data and one or more nodes may play the role of relay nodes (RN) [35]. Multi hop communication is the promising solution to increase network coverage and throughput. Transmission power of the senor nodes can be reduced to transmit the data at the short distance and to reduce the interference among the signals. This is advantageous in terms of spatial reuse of frequency. But a node playing the role of RN can deplete its energy earlier than other nodes so this problem should be examined and tackled by the routing protocols. Many different technologies are under exploration like fixed relays (Relays that are not connected to the backbone of the network), movable relays (Relays, which agree to transmit the packets of each other's) and hybrid relays (Relays, which are fixed but are situated on the body of mobile objects). The use of relay nodes is very beneficial in terms of scheduling, interference management, network lifetime, adaptive modulation etc. Due to advantages of multi hop communication, many researchers have developed relay based routing protocols and in future, it can be considered vital to give attention to short-range communication where power levels of nodes can be controlled. Many protocols falls under the category of multi hop communication. They are categorized and explored in successive chapters.

References

1. Akyildiz F, Kasimoglu IH (2004) Wireless sensor and actor networks: research challenges. Ad Hoc Netw 2(4):351–367
2. Karl H, Andreas W (2007) Protocols and architectures for wireless sensor networks. Wiley, New York
3. Estrin D, Culler D, Pister K, Sukhatme G (2002) Connecting the physical world with pervasive networks. In: IEEE Pervasive Computing, Jan 2002, pp 59–69
4. Stankovic JA, Wood AD, He T (2011) Realistic applications for wireless sensor networks. In: Theoretical Aspects of Distributed Computing in Sensor Networks. Springer, Heidelberg, pp 835–863
5. Oliveira LM, Rodrigues JJ (2011) Wireless sensor networks: a survey on environmental monitoring. J Commun 6(2):143–151

6. Aberg PA, Togawa T, Spelman FA (eds) (2002) Sensors in medicine and healthcare. Wiley, New York

7. Ko J, Lu C, Srivastava MB, Stankovic J, Terzis A, Welsh M (2010) Wireless sensor networks for healthcare. Proc IEEE 98(11):1947–1960

8. Gungor VC, Hancke GP (2009) Industrial wireless sensor networks: Challenges, design principles, and technical approaches. Industr Electron IEEE Trans 56(10):4258–4265

9. Clare LP, Romanov N, Twarowsk A (1999) Wireless sensor networks for area monitoring and integrated vehicle health management applications. American Institute of Aeronautics and Aeronautics (AIAA)

10. De Lara E, LaMarca A, Satyanarayanan M (2008) Location systems: an introduction to the technology behind location awareness. Morgan & Claypool Publishers, San Rafael, p 88. ISBN 978-1-59829-581-8

11. National Research Council (U.S.), Committee on the Future of the Global Positioning System, National Academy of Public Administration (1995) The global positioning system: a shared national asset: recommendations for technical improvements and enhancements. National Academies Press, p 16. ISBN 0-309-05283-1. Retrieved 16 Aug 2013

12. Schiettecatte B (2004) Interaction design for electronic musical interfaces. In Proceedings of the conference on Extended abstract on human factors in computing systems, Apr 2004, Vienna, Austria

13. Younis M, Youssef M, Arisha K (2003) Energy-aware management in cluster-based sensor networks. Comput Netw 43(5):649–668

14. Hou YT, Shi Y, Sherali HD (2005) On energy provisioning and relay node placement for wireless sensor networks. IEEE Trans Wirel Commun 4(5): 2579–2590

15. Oyman EI, Ersoy C (2004) Multiple sink network design problem in large scale wireless sensor networks. In: Proceedings of the IEEE international conference on communications (ICC 2004), Paris, June 2004

16. Khanna R, Liu H, Chen HH (2006) Self-organization of sensor networks using genetic algorithms. In: Proceedings of the 32nd IEEE international conference on communications (ICC'06), Istanbul, Turkey, June 2006

17. Moscibroda T, Wattenhofer R (2005) Maximizing the lifetime of dominating sets. In: Proceedings of the 19th IEEE international parallel and distributed processing symposium (IPDPS'05), Denver, Colorado, Apr 2005

18. Dasgupta K, Kukreja M, Kalpakis K (2003) Topology-aware placement and role assignment for energy-efficient information gathering in sensor networks. In: Proceedings of 8th IEEE symposium on computers, 2003. doi:10.1109/ISCC.2003.1214143

19. Younis O, Fahmy S (2004) HEED: a hybrid, energy-efficient, distributed clustering approach for ad-hoc sensor networks. IEEE Trans Mobile Comput 3:366–379. doi:10.1109/TMC.2004.41

20. Heinzelman WB, Chandrakasan AP, Balakrishnan H (2002) Application specific protocol architecture for wireless micro sensor networks. IEEE Trans Wirel Network, pp 660–670. doi:10.1109/TWC.2002.804190

21. Bandyopadhyay S, Coyle E (2003) An energy efficient hierarchical clustering algorithm for wireless sensor networks. In: Proceedings of the 22nd annual joint conference of the IEEE computer and communications societies (INFOCOM 2003), San Francisco, California, Apr 2003

22. Amis AD, Prakash R, Vuong THP, Huynh DT (2000) Max-Min d-cluster formation in wireless ad hoc networks. In: Proceedings of 20th joint conference of the IEEE computer and communications societies (INFOCOM'2000), Mar 2000. doi:10.1109/INFCOM.2000.832171

23. Garcia F, Solano J, Stojmenovic I (2003) Connectivity based k-hop clustering in wireless networks. Telecommun Syst 22(1):205–220

24. Fernandess Y, Malkhi D (2002) K-clustering in wireless ad hoc networks. In: Proceedings of the 2nd ACM international workshop on principles of mobile computing (POMC'02), Toulouse, France, Oct 2002

25. Younis M, Akkaya K, Kunjithapatham A (2003) Optimization of task allocation in a cluster-based sensor network. In: Proceedings of the 8th IEEE symposium on computers and communications (ISCC'2003), Antalya, Turkey, June 2003
26. Krishnamachari B, Estrin D, Wicker S (2002) Modeling data centric routing in wireless sensor networks. In: Proceedings of IEEE INFOCOM, New York, June 2002
27. Gupta G, Younis M (2003) Fault-tolerant clustering of wireless sensor networks. In: Proceedings of the IEEE wireless communication and networks conference (WCNC 2003), New Orleans, Louisiana, Mar 2003
28. Oyman EI, Ersoy C (2004) Multiple sink network design problem in large scale wireless sensor networks. In: Proceedings of the IEEE international conference on communications (ICC 2004), Paris, June 2004
29. Akyildiz IF et al (2002) A survey on sensor networks. IEEE Commun Mag 40(8):102–111
30. Al-Karaki JN, Kamal AE (2004) Routing techniques in wireless sensor networks: a survey. Wirel Commun IEEE 11(6):6–28
31. Pantazis N, Nikolidakis SA, Vergados DD (2013) Energy-efficient routing protocols in wireless sensor networks: a survey. Commun Surv Tutor IEEE 15(2):551–591
32. Pottie G, Kaiser W (2000) Wireless integrated sensor networks (WINS). Commun ACM 43 (5):51–58
33. Merrill WM, Sohrabi K, Girod L, Elson J, Newberg F, Kaiser W (2002) Open standard development platforms for distributed sensor networks. In: Proceedings of SPIE—unattended ground sensor technologies and applications IV (AeroSense 2002), vol 4743, Orlando, FL, Apr 2002, pp 327–337
34. Hill J, Szewcyk R, Woo A, Culler D, Hollar S, Pister K (2000) System architecture directions for networked sensors. In: Proceedings of 9th international conference on architectural support for programming languages and operating systems (ASPLoS IX), Cambridge, MA, Nov 2000, pp 93–104
35. Broch J, Maltz DA, Johnson DB, Hu YC, Jetcheva J (1998) A performance comparison of multi-hop wireless ad hoc network routing protocols. In: Proceedings of the 4th annual ACM/IEEE international conference on Mobile computing and networking (pp 85–97). ACM

Chapter 2
Multi-hop Energy Efficient Routing

Abstract The very challenging requirement of WSN is energy efficiency, which depends upon the distributed design and dynamic topology of the network. This requirement can be fulfilled by the multi-hop energy efficient routing protocols. The objective of all the energy efficient protocols is to extend the network's lifetime. Sensor nodes are battery operated, so generally concern of all routing protocols is to conserve energy. In this chapter, energy efficient protocols are categorized into two types (i) heterogeneity based energy efficient routing protocols and (ii) chain based energy efficient routing protocols. Comparisons of various energy efficient routing protocols are specified here with open issues. A systematic and comprehensive taxonomy of various energy aware schemes are discussed in depth. This chapter is focused on various energy conservation schemes and hence, discussion of various routing protocols gives the readers a new insight.

Keywords WSN · Multi-hop routing · Energy efficiency · Classification · Comparison · Future trend

2.1 Introduction

A lot of work has been carried out in recent past for wide development of applications in WSN and to support these applications many energy efficient routing protocols have developed to deal with limited energy issue of WSN. Sensors may be used in the area monitoring to locate the object, to measure the humidity, to measure the air pollution, to measure the water density etc. WSN is comprised of many tiny sensor nodes. A tiny sensor node is consisted of three subsystems, a communication unit for data transmission, a sensing unit to capture the data from physical environment and a processing unit to process the gathered data. Power source of sensor nodes is battery operated and it is inconvenient to recharge it when SNs are once deployed in unattended atmosphere. A long network lifetime is required from sensor networks by most of the applications, as few months or

© The Author(s) 2016

S. Rani and S.H. Ahmed, *Multi-hop Routing in Wireless Sensor Networks*,
SpringerBriefs in Electrical and Computer Engineering,
DOI 10.1007/978-981-287-730-7_2

several years. But how to prolong the network lifetime of the WSN with battery operated nodes is the major issue. Energy conservation is the key concern for WSNs. In this chapter, we have explored various multi hop routing protocols which are energy efficient. The energy of the routing protocols is affected by the many factors and some are discussed as below:

(a) Energy Expenditure in One round: One round is defined as the execution of the protocol for one time in which data of all the sensor nodes is transmitted to the BS. This is the total count of the energy expenditure of all the nodes in one round.

(b) Energy Consumption in Data aggregation: It is defined as the depletion of the energy in aggregating data. Some routing protocols are multi-hop and they transmit their data through the CHs and data is aggregated at the CHs by the nodes. So CHs consume energy in this task and it depends upon the condition whether the data was already processed or still have to be processed by the CH.

(c) Residual Energy Threshold: It is the term which makes aware of the remaining energy of the nodes so that it could become easier to know how much load can be handled by the sensor node. Some Routing protocols do not elect the sensor nodes as the CHs when their energy level is less than the threshold.

(d) Lifetime of the Network: It is defined as the time till the death of the last node of the network. Routing protocols use this factor to predict the total time of the execution of the network.

(e) Energy and the Length of the data: length of the packet to be transmitted is the factor, which should be considered while computing the energy consumption as lengthy packets will consume more energy of the nodes than the short packets.

(f) Enhancement of the Reliability at Cost of Energy: Requirement of the applications decide whether the redundancy of the data is required or not. As some applications require that data must be reachable to the BS at any cost the redundancy of data is preferred and same data packets are sent via the multiple paths. In contrast to these applications some require that energy should be conserved and where it is the common tendency to reduce the redundancy.

(g) Sleep and the Wake-up Schedule: In some routing protocols if the data communication is not on the continuous basis then only few nodes are active and other nodes turned off their antennas to save energy. After some time interval they wake up to receive the data buffered or intended for them.

(h) Idle listening or overhearing: Idle listening is the mode when sensor nodes do not send or receive any data but just listen to the channel and overhearing is the process when data is transmitted to some neighbor nodes and other nodes are in the range on their neighbor and overhear the messages which are not intended for them. Both the factors should be decreases by controlling the radio antennas and the range of the nodes.

(i) Total Energy Consumption by the Node: Mainly energy of the node is depleted in the sensing, transmitting, processing and aggregation of the data. So energy cost of the nodes is computed in these terms.

Fig. 2.1 Classification of
multi-hop energy efficient
protocols

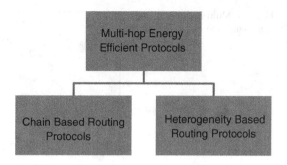

(j) Energy Consumption in Informatics Messages: Time to time some messages
    are transmitted by base station to the nodes to know their status and sensor
    nodes acknowledge back to the BS. These messages should be kept as mini-
    mum as possible to conserve energy.
(k) Travelling Distance of the Data: More energy of the sensor node is consumed
    if the packet has to be transmitted at the long distance that is why some
    protocols use the clustering scheme to reduce the communication distance
    among the nodes.

Factors discussed above account for the computation of the energy metric of the
network. Energy based data routing protocols are as follows. Energy efficient
protocols can be classified into two categories [1] as shown in Fig. 2.1.

2.2 Multi-hop Energy Efficient Routing Protocols

Development of energy efficient protocols is the need of the hour. WSN is the
energy constrained network due to the limited power, low memory, low processing
battery operated sensor nodes [2]. Routing protocols mainly considers the try to
maximize the life of the WSN by developing the optimal shortest routing paths and
by minimizing the data travelling distance among the nodes. To make the WSN
work for several years, other issues are also considered like switching off the radio
components of the nodes whenever they are not in use and nodes have to self
organize to maximize the energy efficiency. The environmental conditions also
affect these nodes where it becomes more essential to take necessary to develop the
new routing algorithms. The main objective of routing is not only to transmit the
data from the source station to destination but to perform this function in energy
efficient way. Architecture and design of the network also affects the energy of the
sensor nodes. To cater the need of the today's WSN, the research in the hardware
also lead some companies (Crossbow, WorldSense, Ember Cooperation etc.) to
develop the ready to use energy efficient sensors which can be deployed directly.
The present research is going on energy efficient routing protocols to optimize the
energy conservation techniques for extending the lifetime of WSN. How to reduce

Fig. 2.2 Multi-hop
communication

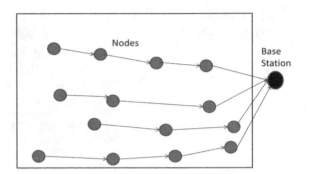

the load of the nodes and processing task; how to reduce the redundant transmissions are the essential factors to be considered and the routing algorithms which will support these, will be preferred in future. The communication cost in terms of energy expenditure sometime increases too much that it lead to the division or the partition of the network.

Factors discussed above account for the computation of the energy metric of the network. Energy based data routing protocols are as follows. Energy efficient protocols can be classified into two categories [1] as shown in Fig. 2.1:

Reduction in delay and distance can be achieved by multi hop transmission (Fig. 2.2), which will increase the reliability of the data transmission and network lifetime. Travelling distance directly affects the energy of nodes so by reducing the distance travelled by the nodes, enhance the energy efficiency of the nodes. Several routing multi hop protocols have been proposed in literature.

In this section, focus is on discussion of chain based data transmission protocols (Fig. 2.3). Main benefit of this scheme is energy conservation by forming the chain between the nodes so that travelling distance of the nodes could be reduced. But problem with chain-based protocols is delay. Few algorithms are working on this metric called energy delay metric.

2.2.1 Chain Based Data Transmission

An Energy-Efficient Unequal Clustering Mechanism for Wireless Sensor Networks (EEUC) is the energy efficient homogeneous multi-hop routing protocol. To compute the energy expenditure, multipath and free space models are used for the periodic data gathering. Clusters near the BS are smaller than the clusters, which are away from the BS. This strategy is used to avoid the problem of hot spot near the BS due to data forwarding. Node's competition radius should be less when its distance from the BS is less. The maximum competition radius is predefined. Each CH maintains the list of adjacent tentative CHs. After the selection of CHs, each node joins the closest cluster with largest received signal strength. Intra cluster routing is same as used in LEACH [3] that is with the help of election of optimal

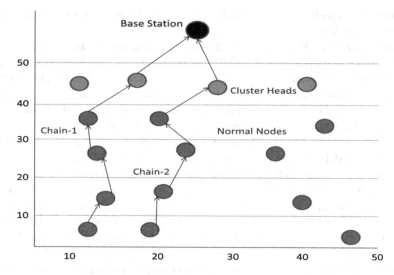

Fig. 2.3 Chain based communication

number of clusters and then role of the CH is rotated among the nodes. For inter cluster communication the distance of the nodes from the BS is calculated, if the distance is less than the predefined threshold distance then the data is transmitted directly, otherwise a forwarding node is find out to forward the data. It is the node from the list of the CHs maintained by the CH, which wants to transmit the data for node. Node with high residual energy is elected as the forwarding node. It has shown its validity over LEACH and HEED but as it has assumed that it does not require the use of GPS then the location of nodes is not known exactly and distance calculation will generate the error which may cause the loss of the packets.

A Hybrid Adaptive Routing Protocol for Mobile Ad Hoc Networks (SHARP) [4] is proposed by Ramasubramanian et al. (2003) to extend the network lifetime. Number of hops, link quality and battery availability; metrics were used in it. To balance the load over the nodes the traffic is distributed on the different possible routes. Topology is also maintained in it to ensure the reliable data transmission. The nodes with received signal strength (RSS) below the threshold values are not included in the routing table. To select the best routes SHARP uses link quality indicator and RSSI. It validated its performance in terms of energy consumption and network lifetime. The routing strategy used in SHARP is complex and it is not easy to compute the hybrid of reactive and proactive which makes it less reliable. A review paper on clustering and routing techniques is proposed in [5]. In this survey, some protocols are described which use multi-hop communication like LEACH-Ensuring the Reliable Data Delivery (LEACH-ER) and it is based on the scheme in which packets exchanged between the nodes are decreased and hence results in the increase of the network lifetime. LEACH-Trust Transmission Mechanism (LEACH-TM) was proposed in 2009 [6]. It considers the residual energy and hop count to make it reliable for data transmission. The distance of the

CH from BS is finding out by RSS (Received signal strength). It has optimized the lifetime of the WSN.

Gateway-Based Energy-Aware Multi-Hop Routing Protocol for WSNs (M-GEAR) was proposed by Nadeem et al. in 2013 [7] in which network has been divided into four regions for energy conservation. Different communication hierarchies are used for communication. Nodes in one region communicate directly to the BS and in the second region they communicate with gateway node. In the third region, clustering hierarchy is used to communicate and in the fourth region, data is passed to the gateway node by the CHs. The number of cluster heads is based on the formula as devised in LEACH. A gateway node is the node located in the centre of the area and its main aim is to forward to the BS which is far away from the area. Nodes near to the gateway transmit data to it and nodes near to the BS, transmit data to the BS directly and other nodes transmit the data to the gateway node via CHs and use multi-hop communication. It has shown its improvement over LEACH in terms of network lifetime.

Hybrid DEEC-Towards Efficient Energy Utilization in Wireless Sensor Networks [8] shows the improvement over DEEC [9], hybrid DEEC has been proposed in which nodes are divided into two categories where 90 % nodes are the normal nodes and 10 % nodes act as the beta nodes (nodes with high residual energy) which collect data from the CHs. Chain of nodes is formed like PEGASIS [10] and the node at the end with less distance to the BS is elected as the beta node. Multi Edged DEEC has shown improvement over H-DEEC. The multipath model computes distance parameter by using the distance of CH to the BS and the average distance of the nodes to the BS. The node with minimum weight will be elected as the leader of the nodes. Localization and interference problems are left to resolve in this protocol. MODLEACH-A Variant of LEACH for WSNs was proposed by Mahmood et al. [11]. It also proposes that for the intra-cluster communication the power level of the nodes should be 10 times less than the power levels used in the inter-cluster communication. Dual power transmission level and modification in election of the CHs improved the network lifetime and throughput.

Dynamic Cluster-based Routing for Wireless Sensor Networks (DCBR) [12] is based on the division of the area into clusters according to K-medoids method. f the value of K is 1 then it behaves like LEACH. Its algorithm divides the area into sub-square areas. Each sub-square has some fixed length and centre C(i, j) where 1 <= i and j <= k. Role of CH is exchanged with the node with high residual energy according to its distance from the BS. It has shown its improvement over LEACH. But by forwarding the data to the one subarea by other subareas will cause that subarea to lose all its nodes fast and in future further information from that subarea will not be available.

Trust-Aware and Low Energy Consumption Security Topology Protocol of Wireless Sensor Network (TLES) [13]: It takes into account the node's residual energy, node's degrees and its distance from the BS to elect the forwarding node in routing. It is divided into two parts, in the first part the trust value is calculated and in the second part the forwarding node is elected. Evaluating node i monitors the quantity sending of the evaluated node j. If the number is lower than the lower limit threshold T_L, the node can be considered as a selfish node. If the number is more than the upper limit threshold T_H, the node may have performed attack as behavior of denial of service. To prevent the forget packets, consistency factor is evaluated which is called analysis of spatial coherence. The receiving node compares the data of the other node with its own data and if the received data is co-related then the data is consistent. The credibility of the node is checked when data is transmitted via relay node. If the computed value is 0 then it is the abnormal node and if its value is close to 1 then this node is reliable node. This computed value if direct trust value but indirect trust value must be considered in multi-hop communication. It is not suitable for the real time applications as it takes long time in searching the path.

Chen et al. proposed an energy efficient Chain Based Hierarchical Routing Protocol, named as CHIRON in 2009 [14]. The main objective of CHIRON was to split the sensing field into a number of smaller areas, so that it can create multiple shorter chains to reduce the data transmission delay and redundant path, and in this way gained the energy conservation per node and hence resulted in prolonging the network life-time. It shows about 15 and 168 % improvements on average data propagation delay, 30 and 65 % improvements on redundant transmission path, respectively over enhanced PEGASIS [15] and PEGASIS protocols respectively. Similarly, an improved energy-efficient PEGASIS-based protocol (IEEPB) was proposed in 2011 by Sen et al. [16]. It shows better performance than EEPB in terms of balancing energy consumption and prolonging lifetime of Wireless Sensor Networks (WSN). MIEEPB was proposed in 2013 by Jafri et al. to improve the delay factor [17]. Sink mobility is fixed in particular location and sink stays in that location. Nodes search location for some time, which is known as sojourn time and gathers the data of all the chain leaders from that region (sojourn region). It has shown its improvement over IEEPB.

The hybrid of LEACH and PEGASIS is Chain Based Mixed Routing (CCM) [18] but limitation of this work is that energy consumption increases with the increase in the network size. The main objective of this protocol is to achieve energy efficiency and it tries to find out the chains, which consume less energy and take less time. Selection of the chain leader can be optimized to enhance the present algorithm. In the same year another protocol, Chain routing with even energy consumption (CREEC) was proposed [19], which outperforms in terms of energy savings and load balancing, as compared to LEACH, PEGASIS and power efficient data gathering and aggregation protocol (PEDAP) [20]. It performs very well for all the WSN sizes and base station distances. EBRAMS [21] has shown better per-formance than LEACH and PEGASIS in delay parameter. Mahajan et al. proposed chain-based protocol in 2013 to form the chain on the basis of the selection value (SV) [22]. But, this protocol does not suit the high dense network as it will lead to more comparisons and will consume more energy of the node. It can affect both the network lifetime and the delay factor.

2.2.2 Heterogeneity-Based Protocols

Many routing protocols assume that sensors have the same capabilities in commu-nication, power, processing, storage and sensing. These protocols are known as

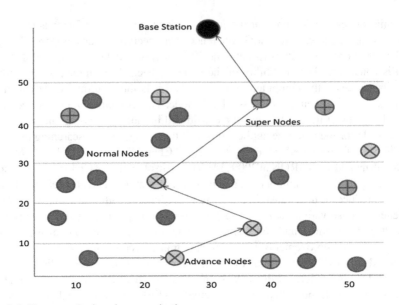

Fig. 2.4 Heterogeneity based communication

homogeneous where links between any two nodes are symmetric. However, in some application sensors with different capabilities are required which are known as heterogeneous. Sensors with same physical properties and using the same platform models are not feasible. For the specific requirements of the sensing applications, heterogeneous nodes are used as shown in Fig. 2.4. iPAQ and CrossBow Mica motes can be used together in the same network as iPAQ motes use more power but can compute the data quickly, so they can be used for the data fusion and CrossBow motes can be used for the sensing as they use very little power and can perform the complex processing. This type of deployment was proposed by the authors of [23]. Using some powerful nodes in the network can extend network lifetime. Inexpensive nodes can be used sensing and powerful nodes can be used for in-network processing. Three layers architecture was proposed in 2005 for heterogeneous network. This architecture is based on the rationale that more energy is consumed in forwarding the data of many nodes by one node. But there should be the optimal number of the powerful nodes at the second layer. With the use of line powered sensor nodes the network lifetime of WSN can be increased. Some sensors are required to provide high quality links to the sink (Highly reliable links at long distance); as provided in 802.11 type connectivity. These links reduces the number of hops in data transmission to the sink. These links are also known as back haul links, which helps in reducing the end-to-end delay. But advantage of this type of heterogeneity depends upon backhaul sensors and their locations. If each sensor is one hop away from the sink or from backhaul sensor then end-to-end success rate is approximately is same as link success rate. Maximum benefit of heterogeneity either link heterogeneity or energy heterogeneity depends upon the size, architecture and density of the network. The length of the shortest path from location (i, j) to the sink that is adjacent to the midpoint of the edge is computed. If backhaul sensor k, is consider one hop away then the distance from that node is calculated. Link heterogeneity can help in maintaining the lifetime of the network.

Gathering Algorithms in Sensor Networks Using Energy Metrics [24]: Main problem in heterogeneous networks is to minimize the detection latency and energy consumption and maximize the information gain for target tracking and localization with dynamic data routing and querying. There are certain events in the network, which should be reported in time so only some sensors required to be active to conserve energy. But election of the active sensors depends upon the important information they have which is balanced by their communication cost. Useful information is defined by the time and space of the events. If X represents the target position of estimation then representation of the current posteriori distribution (belief) with given set of measurements Z1, Z2...Zn is calculated. Belief is based on the measurements of the sensor nodes and so it is required to minimize the communication to select the best belief. Chu et al. proposed routing protocol for heterogeneous sensor network in 2002 [25]. Two novel techniques, constrained anisotropic diffusion routing (CADR) and information-driven sensor querying (IDSQ) were used for energy-efficient routing and data querying in ad hoc sensor networks. It proved its optimized results in detection latency and node failures. To derive the sensor measurement to belief state information driven sensor querying protocol (IDSQ) aims at optimizing the sensor selection. A sensor l is elected as the leader from the N set of sensors which will be responsible for electing the optimal sensors based on some information like geometric measure given as the Mahalanobis distance from the destination to the sensor node under consideration queries for data. The leader must have the knowledge of certain features as locations. Each sensor will wait for the query from the leader and after queried, each sensor sends the information back to the sender. When the target is in the range of the cluster nodes then leader is activated and computes the belief using its own dimension. Based on, its quality the leader may stop processing or may continue with sensor selection. If belief is not good then it runs its own algorithm for sensor selection then it updates it belief state by incorporating new sensors. It runs again if the selection is again not good. In this way, it helps in conservation of the energy of the nodes if they do not have useful information but at the cost of less reliability. RSVP [23] is based on notion of the heterogeneous receivers [26]. This protocol is not scalable and resources reservation scheme is static. Resources should be assigned dynamically so that few resources could cater the need of all the nodes. Supporting Hierarchy and Heterogeneous Interfaces in Multi-Hop Wireless Ad Hoc Networks [27] has the same problem of addressing which arises when two nodes handled by different administrators want to communicate [28]. Delay and energy problems are encountered in the heterogeneous protocol [26] developed for dense network because nodes have to broadcast the message again and again to claim itself as the CH and it has no support for recovery from fault [29]. In SEP [30] routing protocol sink is assumed in centre but it does not present the good idea for the applications where BS is located at corner and it is not scalable. For scalability purpose [31], a routing scheme with heterogeneous sensor nodes was proposed. It shows its improvement over Directed diffusion (DD) and Granted delivery (SWR) protocols. In it data is forwarded to the BS by the relay nodes. Forwarding data by relay nodes will cause the relay nodes to lose their energy faster than other nodes that can make network obsolete and for high dense applications it is not practicable.

Energy efficient heterogeneous clustered scheme [32] uses three types of heterogeneity that are energy heterogeneity, computational heterogeneity and link heterogeneity. Without energy heterogeneity the link and computational can't bring the positive results in terms of QoS metrics of WSN. This represents the same structure as used in SEP but with minor difference that in SEP only one level of energy heterogeneity but it uses tow levels of heterogeneity. But it has not explained two concepts (1) the reliability and (2) how the network will cope- up with the nodes after the failure of some nodes. Chan et al. proposed the other protocol, WSN, in the same year, which is named as a geography-based heterogeneous hierarchy routing (GBHHR) protocol [33]. A dormancy system is adopted in this protocol for heterogeneous nodes to solve the problem of election of CH to conserve energy in better way than traditional protocols. In 2014, a routing protocol based on topologies for heterogeneous wireless sensor networks (ROUT) was proposed [34]. It follows the idea of division of nodes into H-sensors (nodes with higher hardware capabilities) and L-sensors (nodes with lower hardware capabilities). How to maintain heterogeneity in the WSN to optimize the QoS metric is the main concern of all the protocols.

2.3 Comparative Analysis

In Table 2.1, we can clearly observe difference in the parametric values of all the energy efficient protocols. All protocols have different measures. As EEUC, LEACH-ER, LEACH-TM, M-GEAR, Hybrid DEEC etc. are less scalable as compared to SHARP, IDSQ/CADR, CHR, GBHHR etc. protocols. Due to load balancing few protocols are more energy efficient than other protocols. Focusing the energy parameter develops these protocols. In this chapter, we have surveyed the many approaches to energy conservation techniques in wireless sensor networks. Main objective is to discuss the comprehensive and systematic categorization of the solutions proposed in the literature.

Table 2.1 Comparative analytical table for multi-hop energy efficient protocols

Protocol	Classification	Scalability	Data aggregation	Energy efficient	Failure recovery	Network type	Load balancing	Latency	Reliability	Mobility	Location awareness
EEUC(2005)	Multi-hop	Low	Yes	Good	No	Homogeneous	Less	Average	Less	Stationery	No
SHARP(2009)	Multi-hop	Good	Yes	Average	Yes	Homogeneous	Yes	High	Yes	Mobile	No
LEACH-ER (2010)	Multi-hop	Low	Yes	Good	No	Homogeneous	No	High	Yes	Stationery	No
LEACH-TM (2009)	Multi-hop	Low	Yes	Good	No	Homogeneous	No	High	Yes	Stationery	No
M-GEAR (2013)	Multi-hop	Low	Yes	Good	No	Homogeneous	No	Average	less	Stationery	No
Hybrid DEEC (2013)	Multi-hop	Low	Yes	Good	No	heterogeneous	No	Average	Average	Average	No
MOD-LEACH (2013)	Multi-hop	Low	Yes	Good	No	Homogeneous	No	Average	Less	Stationery	No
DCBR(2014)	Multi-hop	Low	yes	Average	No	Homogeneous	No	Average	Less	Stationery	Yes
TLES(2014)	Multi-hop	Average	Yes	Good	No	Homogeneous	No	Average	Yes	Stationery	Yes
HEED(2004)	Multi-hop	Low	Yes	Average	Less	heterogeneous	Yes	Average	Less	Stationery	No
CCRP(2008)	Chain- Based	Low	Yes	Very Good	No	Homogeneous	No	Very High	Less	Stationery	Yes
CHIRON (2009)	Chain- Based	Average	Yes	Good	No	heterogeneous	No	Low	Less	Stationery	No
IEEPB(2011)	Chain- Based	Low	Yes	Average	No	Homogeneous	No	Average	Less	Sink Mobile	No
MIEEPB (2013)	Chain- Based	Low	Yes	Average	No	Homogeneous	No	Average	Less	Sink Mobility in Particular Locations	No

(continued)

Table 2.1 (continued)

Protocol	Classification	Scalability	Data aggregation	Energy efficient	Failure recovery	Network type	Load balancing	Latency	Reliability	Mobility	Location awareness
IECBSN (2013)	Chain-Based	Very Low	Yes	Low	No	Homogeneous	No	Very High	Very Less	Stationery	No
IDSQ/CADR (2002)	Heterogeneous	High	Less	Good	No	Heterogeneous	No	Low	Average	Nodes are Mobile	Yes
AODV-DSDV (2002)	Heterogeneous	High	Less	Good	No	Heterogeneous	No	Low	Average	Nodes are Mobile	No
CHR(2005)	Heterogeneous	Average	Yes	Good	No	Heterogeneous	No	Low	Less	Stationery	Yes
GBHHR (2008)	Heterogeneous	High	Yes	Good	No	Heterogeneous	No	Average	Average	Stationery	Yes
LayHet(2012)	Heterogeneous	Average	Less	Very Good	No	Heterogeneous	No	Low	Average	Stationery	No
ROUT(2014)	Heterogeneous	High	Yes	Depends upon the topology used	No	Heterogeneous	No	Depends upon the topology used	Average	–	Yes

2.4 Summary and Future Trends

A large number of energy efficient protocols have been proposed in recent past. Still there is lot of work to be done and research is continued. We have stressed on many approaches, like data driven and mobility based ideas. Final observations can be drawn on the basis of different techniques of energy management. An increasing interest towards sparse network architecture can be noticed but such network can be useful only if the advantages of mobility are exploited. The dependence and complexity upon the collaborative efforts of WSN require use of energy efficient protocols through which connectivity of the network can be maintained. New protocols are required to improve network lifetime, delay and network connectivity.

References

1. Pantazis N, Nikolidakis SA, Vergados DD (2013) Energy-efficient routing protocols in wireless sensor networks: a survey. IEEE Commun Surv Tutorials 15(2):551–591
2. Li C, Ye M, Chen G (2005) An energy-efficient unequal clustering mechanism for wireless sensor networks. IEEE Int Conf Mobile Adhoc Sens Syst Conf. doi:10.1109/MAHSS.2005. 1542849
3. Heinzelman WB, Chandrakasan AP, Balakrishnan H (2002) Application specific protocol architecture for wireless micro sensor networks. IEEE Trans Wirel Netw 660–670. doi:10. 1109/TWC.2002.804190
4. Ramasubramanian V, Haas ZJ, Sirer EG (2003) SHARP: a hybrid adaptive routing protocol for mobile ad hoc networks. In: Proceedings of the 4th ACM international symposium on mobile ad hoc networking and computing (MobiHoc '03). ACM, New York, NY, USA, pp 303–314. doi:10.1145/778415.778450
5. Tyagi S, Kumar N (2013) A systematic review on clustering and routing techniques based upon LEACH protocol for wireless sensor networks. J Comput Netw Appl 623–645. doi:10. 1016/j.jnca.2012.12.001
6. Weichao W, Fei D, Qijian X (2009) An improvement of LEACH routing protocol based on trust for wireless sensor networks. In: 5th international conference on wireless communications, networking and mobile computing, 2009. WiCom'09. IEEE
7. Nadeem Q et al. (2013) M-GEAR: gateway-based energy-aware multi-hop routing protocol for WSNs. In: 2013 8th international conference on broadband and wireless computing, communication and applications (BWCCA). IEEE. doi:10.1109/BWCCA.2013.35
8. Qing L, Qingxin Z, Mingwen W (2006) Design of a distributed energy-efficient clustering algorithm for heterogeneous wireless sensor networks. Comput Commun 29(12):2230–2237
9. Khan MY et al (2013) Hybrid DEEC: towards efficient energy utilization in wireless sensor networks. arXiv preprint arXiv:1303.4679
10. Lindsey S, Raghavendra CS (2002) PEGASIS: power-efficient gathering in sensor information systems. In: Aerospace conference proceedings, 2002. IEEE, vol 3. IEEE
11. Mahmood D et al (2013) MODLEACH: a variant of LEACH for WSNs. In: 2013 8th international conference on broadband and wireless computing, communication and applications (BWCCA). IEEE. doi:10.1109/BWCCA.2013.34
12. Zhao L, Chen Z, Sun G (2014) Dynamic cluster-based routing for wireless sensor networks. J Netw 9(11):2951–2956. doi:10.4304/jnw.9.11.2951-2956
13. Chen Z, He M, Liang W, Chen K. Trust-Aware and low energy consumption security topology protocol of wireless sensor network. J Sens, Article ID 716468 (in press)

14. Chen KH, Huang JM, Hsiao CC (2009) CHIRON: an energy-efficient chain-based hierarchical routing protocol in wireless sensor networks. In: Wireless telecommunications symposium, 2009. WTS 2009. IEEE

15. Yueyang L, Hong J, Guangxin Y (2006) An energy-efficient PEGASIS-based enhanced algorithm in wireless sensor networks. China Commun 91–97

16. Sen F, Bing Q, Liangrui T (2011) An improved energy-efficient pegasis-based protocol in wireless sensor networks. In: IEEE 8th international conference On fuzzy systems and knowledge discovery (FSKD), vol 4. IEEE, pp 2230–2233

17. Jafri MR et al (2013) Maximizing the lifetime of multi-chain pegasis using sink mobility. arXiv preprint arXiv:1303.4347

18. Tang F, You I, Guo S, Guo M, Ma Y (2012) A chain-cluster based routing algorithm for wireless sensor networks. J Intell Manuf 23(4):1305–1313

19. Shin J, Sun C (2011) CREEC: chain routing with even energy consumption. J Commun Netw 13(1):17–25

20. Tan HÖ, Körpeoğlu I (2003) Power efficient data gathering and aggregation in wireless sensor networks. ACM Sigmod Record 32(4):66–71

21. Yarvis M, Kushalnagar N, Singh H, Rangarajan A, Liu Y, Singh S (2005) Exploiting heterogeneity in sensor networks. In: Proceedings IEEE INFOCOM'05, vol 2, Miami, FL, pp 878–890

22. Mahajan S, Malhotra J, Sharma S (2013) Improved enhanced chain based energy efficient wireless sensor network. Wirel Sens Netw 5(4):84–89. doi:10.4236/wsn.2013.54011

23. Zhang L et al (1993) RSVP: a new resource reservation protocol. Netw IEEE 7(5):8–18

24. Lindsey S, Raghavendra C, Sivalingam KM (2002) Data gathering algorithms in sensor networks using energy metrics. IEEE Trans Parallel Distrib Syst 13(9):924–935

25. Chu M, Haussecker H, Zhao F (2002) Scalable information-driven sensor querying and routing for ad hoc heterogeneous sensor networks. Int J High Perform Comput Appl 16(3):293–313

26. Xu K, Gerla M (2002) A heterogeneous routing protocol based on a new stable clustering scheme. MILCOM 2002. Proceedings, vol 2. IEEE

27. Broch J, Maltz DA, Johnson DB (1999) Supporting hierarchy and heterogeneous interfaces in multi-hop wireless ad hoc networks. In: 4th international symposium on parallel architectures, algorithms, and networks, 1999. (I-SPAN'99) Proceedings. IEEE

28. Smaragdakis G, Matta I, Bestavros A (2004) SEP: a stable election protocol for clustered heterogeneous wireless sensor networks. Boston University Computer Science Department

29. Du X, Lin F (2005) Improving routing in sensor networks with heterogeneous sensor nodes. In: Vehicular technology conference, 2005. VTC 2005-Spring. 2005 IEEE 61st, vol 4. IEEE

30. Qing L, Zhu Q, Wang M (2006) Design of a distributed energy-efficient clustering algorithm for heterogeneous wireless sensor networks. Comput Commun 29(12):2230–2237

31. Kumar D, Aseri TC, Patel RB (2009) EEHC: energy efficient heterogeneous clustered scheme for wireless sensor networks. Comput Commun 32(4):662–667

32. Kumar S, Prateek M, Bhushan B (2014) Energy efficient (EECP) clustered protocol for heterogeneous wireless sensor network. arXiv preprint arXiv:1408.3202

33. Chen X et al (2008) A geography—based heterogeneous hierarchy routing protocol for wireless sensor networks. In: 10th IEEE international conference on high performance computing and communications, 2008. HPCC'08. IEEE

34. Ludovico Guidoni, Daniel et al (2014) RouT: a routing protocol based on topologies for heterogeneous wireless sensor networks. Latin Am Trans IEEE (Revista IEEE America Latina) 12(4):812–817

Chapter 3
Multi-hop Reliability and Network Operation Routing

Abstract Many routing, power aware and data diffusion protocols have been designed especially for WSN, which are reliability and network operation based. The presentation of various ideas on multi-hop reliability and network operation based routing protocols is given here. Unattended sensor nodes deployed randomly in the network area require reliable routing. In recent years, an extensive research has addressed various issues for the coordination and management of nodes for efficient operations and efficient routing. The mission of maintaining and finding routes is nontrivial as sudden changes in status of nodes lead to unpredictable consequences. In this chapter, comparative study of various routing protocols has been presented to pave new ways to researchers. In this chapter, we have classified reliability and network operation routing in to five types (i) coherent based, (ii) QoS based, (iii) multipath based, (iv) query based and negotiation based.

Keywords Multi-hop routing · Reliability · Network operations · Query · Quality of service (QoS) · Coherent

3.1 Brief History

Designing of Routing protocols for demands of existing and developing applications is an important issue. Multi-hop routing protocols are classified into five categories under network operation and reliability (Fig. 3.1). Many protocols are proposed in literature based on single path communication without considering the increased load on that path. Routing through single path requires less computation but it reduces the network throughout [1, 2]. In case of the single path failure, new routes need to be searched, which can waste the resources and energy of the nodes. Due to resource constraints and limitations of single path routing, multipath routing became the promising solution to exploit the benefits of WSN. Wireless sensor network is different from other networks in operation as it needs some global address scheme to identify the location of the sensor nodes for communication. In

Fig. 3.1 Multi-hop reliability and network operation based routing protocols

the multi-hop network, the problems like fading and error rate affect the maneuver of the protocol. QoS-aware and coherent based protocols are proposed in recent past which consider end-to end delay requirements while setting up the paths in the sensor network. Data processing is also one of the major concerns of WSN. If raw data is forwarded without any processing to the aggregating node (AN) (non-coherent technique), it increases the complexity of that node. In coherent routing protocols, nodes forward the data after minimum processing and cooperate with each other in load sharing. This technique is helpful in long network lifetime. Transmission of redundant data also wastes the resources and time so some protocols are proposed to eliminate redundant data transmission. The enthusiasm behind this literature work is that flooding will overlap and implosion the transmitted data, then nodes will have copies of duplicate data. Data can also be transmitted on the basis of queries made by nodes. In this type of routing protocols, destinations nodes send their queries in the network, and nodes which have data queried by the nodes, reply back with query. Data is forwarded via shortest routes to conserve energy of the nodes.

Several challenges in the way of WSN can be summarized as below:

(a) Area Coverage: This is the one indispensable problem of the WSN. Every corner of the area should be reachable by the network so that information of each cell of the area could be collected. It is very important issue in many applications as military area surveillance, border security surveillance, environmental monitoring etc.

(b) Scalability and Time varying Features: WSN ranges from small area with few sensor nodes to the larger areas with more than thousand nodes. Routing algorithms should take care of the scalability that is the algorithms which work for the fewer number of nodes should also work for the larger areas. Due to the dynamic behavior and adaptive nature of the sensor nodes they must be self configurable and self organized to perform the current activities. With the novel algorithms sensor nodes can achieve the good performance even in the case of environmental interference and noise errors.

(c) Energy, Computation and Communication: These features affect the QoS metrics of the WSN. For the provisioning of these metrics, routing protocols should consider the bandwidth, power supply, battery power and processing load of the network.

(d) Resource Awareness: Routing protocols which do not consider the available resources for the communication and routing, they cannot cope up with QoS metrics. Reliability and the quality of the required attributed cannot be compromised for the long network lifetime of the WSN.

(e) Overlapping of the Zones: Sometimes sensor nodes gather the data of the overlapped geographical areas and transmit it to the common neighbor node, which is the wastage of the bandwidth and energy of the nodes. New protocols need to be developed to tackle with this problem.

(f) Traffic Implosion: Nodes gather data and forward it to the other nodes to make them aware of the various events or they respond to the queries of the source node. But same data is transmitted to the source nodes by the multiple nodes. It causes the traffic implosion problem. This problem can be solved by the careful selection of the boundaries of the nodes, which will respond to the queries of the source node. Local optimization routing algorithms can help in this problem.

(g) Single base station or the sink: In some protocols single base station or the sink is assumed to gather the data from the multiple nodes. To enhance the QoS metrics network area can be partitioned into different levels and each level may have sink to collect the data which will directly affect the distance travelled by node's packets.

(h) Data Pattern: Some unpredictable events may cause the periodic data to be generated and in some applications periodic real time data is sensed and monitored. These periodic and periodic event timings are generally either predictable or not known prior. Hybrid data handling pose a new obstacle in front of WSN. It requires routing protocols which can differentiate between both types of data.

(i) Unreliable Medium of Communication: WSN operated in the unreliable medium of radio waves. This medium is less reliable due to the environmental factors, cross interference channels and noise errors. Selection of the routing algorithms for particular application considers this factor so that communication should be continued even in the case of failure of some nodes. Due to above challenge the routing protocols are either application specific or geographical or network based.

3.2 Reliability and Network Operation Based Protocols

Few state of the art protocols based on the operation of the routing protocols and reliability are discussed below.

3.2.1 Multipath-Based Protocols

Data transmission between the sensor nodes and the sink can take place by two routing designs, which are single path routing and multipath routing. In single path routing data by sensor nodes is transmitted to the sink by shortest path. In multipath routing data is transmitted by finding the multiple shortest paths to the sink for even distribution of load on all paths. Disjoint multipath protocol was proposed by Ganesan et al. [3]. In this routing the path, which is best presented is known as primary path while others are less advantageous as they take longer time than primary path. It requires the global knowledge of network topology by all the nodes but this procedure can be implemented even at local level. A sink can examine which of its neighbor node can provide premier quality data with less delay as done in directed diffusion that is followed by sink's next neighbor to from the primary path. The same operation is repeated by the sink for the next neighbor to generate the alternate path but it is not guaranteed here that the alternate paths find out by this procedure will be same as can be searched by knowledge of global topology. In this routing, the primary path is computed by source node, which does not include the node. Best alternate paths may or may not be disjoint from the primary path, are called braided multipath (Fig. 3.2). Resilience and maintenance overhead, two metrics are used to evaluate the performance of two novel algorithms.

It was proved in this paper that lower maintenance overhead is required in idealized multipath than 2-disjoint idealized Multipath. It was also shown that multipath with lower latency can't be computed at local level and braided multipath has 50% high resilience than disjoint multipath.

In MMSPEED [4] probabilistic QoS guarantee in WSN was proposed for reliability and timeliness. It proved that end-to-end reliability can be achieved by multipath forwarding. This method is implemented at local level without global information of network state. Lu et al. proposed the distributed multipath routing protocol for WSN in 2007. Its main objective is to balance load on the network by

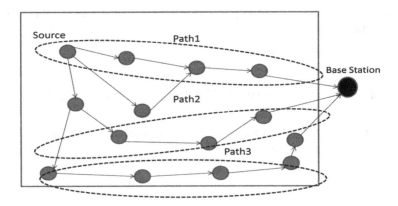

Fig. 3.2 Multipath communication

optimally allocating traffic rate on each path. An energy-efficient multipath routing protocol for wireless sensor networks was proposed by Lu et al. [5]. With the help of load balancing and multipath routing algorithm, it proved its validity over directed diffusion, direct transmission, N to 1 routing and energy aware protocols. It used the data aggregation and load balancing algorithm of [6] and [7] respectively. To measure the load balance transmission on the nodes, Chebyshev sum inequality is used in load balancing algorithm. Basically this protocol has assumed to allocate an optimal traffic rate to each node. It has shown the improvement in the parameters of energy efficiency, delay and control overhead. The aim of MCMP [8] is to provide the soft quality of metrics to the packets being routed. It suggested that E2E is not suitable scheme to estimate the reliability in the large-scale network as its complexity increases with network and it does not give the proper estimate. So it has considered the intermediate nodes to estimate the path with more reliable factor. REER [9] has used the multiple cooperative nodes [CNs] to transmit the data to the sink reference nodes [RNs], which are selected between the sink, and the node. It is developed for the high-density networks. Multipath algorithm for video streaming was considered best as opposed to the direct diffusion [10]. It improved itself better than EDGE protocol in terms of throughput. With same objectives an energy efficient and QoS based multi-path hierarchical routing protocol [11] was proposed. Residual energy, remaining buffer size, signal-to-noise ratio and distance to sink, parameters is used for path discovery and CH election. Bezier-based multipath routing algorithm [12] outperformed over SWEEP and other shortest path schemes in term of delay and network lifetime. Multipath routing with topology control has been proposed by Iqbal et al. [13] to manage interference. Non-coordinated inter-ference enslavement has been removed by this scheme, which resulted in, little number of hops, and end-to-end throughput has been improved.

Based on the above-mentioned protocols it is concluded that multipath protocols can be categorized as N to 1 multipath discovery, Braided and Disjoint multipath based routing schemes. In disjoint, no node is common among the paths and alternate paths are not dependent on the primary path. This leads to more latency in transmitting the data to the sink in case if alternate path replace the primary path. In braided scheme, alternate paths are not disjoint from the primary path. In both of these schemes multiple disjoint paths are find out between the sink and the node. Third method N to 1, is based on simple flooding to find out the multiple disjoint paths. Main objectives of all these protocols are to attain maximum energy con-servation, delay tolerance and reliability for WSN.

3.2.2 Query Based Protocols

In some applications the nodes respond only when data request is send to them. For example if node *A1* needs some data *d*, it will forward its request (query) to the neighbor node *A2* which will be entertained by that node in case availability of the same data (response) otherwise it can send report to the node *A1* with negative message. It is also possible that query is broadcasted to all the nodes and node having the data of that query replies back. Many protocols fall under this category (Fig. 3.3), some of them are Directed Diffusion [14], TEEN [15], Enhanced APTEEN [16] etc. A message interest is a query, which describes the need of the user. Physical phenomenon is portrayed by the data and sensed phenomena by event and gradient describes the direction state where each node receives interest. Gradients are attributed by data rate, interval and expiration time. Tasks are names designed with the attribute and value pairs. It proposed the reactive routing techniques. Routes are established on demand only. Diffusion present in this protocol is different from the traditional network as it is centric and data communication is neighbor to neighbor. They do not require global identifiers. It is shown that it is different from ad-hoc reactive routing scheme as it uses directional flooding and redundant messages are sent on the multiple paths. After that this multiplicity is reduced based on the route performance. The naming technique of this protocol is used to save the energy by avoiding unnecessary operations of routing layer. When a route between node and the sink fails, then new path is identified by the reinforcement method that is done by searching the path between the nodes operating at the lower speeds. But this protocol is application specific as it works for the on demand query and is not suitable for the applications with the requirement of continuous transmission of data. Naming design is also application dependent, which should be known in advance. To match the query with data also requires extra control overheads and needs more processing. In some query based routing schemes, query is broadcasted to all the nodes and it works like flooding. But when data required by nodes is at small scale then flooding is not

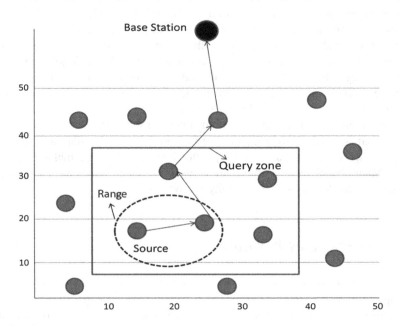

Fig. 3.3 Query routing

the good approach. So events, which are less in number, can be flooded rather than query that are large in number [17]. In this case query will be forwarded to the nodes, which have sensed some events. When some node detects some event then it is stored in the event table and an agent is generated which propagates information about local events to distant nodes. If the node has information of route of event then query will be transmitted on that path otherwise it is forwarded in the random direction until it reaches the target location. It proved better in the cases where geographical routing cannot be established. It shows its validation over event flooding. But overhead of generating of the agent during an event require more processing at the node's end. Then transmission of information by the agent again increases the load that makes it more complex. If the number of nodes is increased, then the chances of retransmission and complexity will increase, so it is not applicable on the large number of nodes.

For declarative querying of sensor network COUGAR approach is introduced [18]. In this approach introduction of query proxy layer at each node for the in-network processing; is used that is the distributed database approach for the WSN. It provides methods of handling data query independent from the network layer. But by introducing the query layer at each node introduces more overhead at each node leading to the requirement of extra processing at each node. Maintenance and synchronization of leader nodes again poses difficulty in time critical applications. Enhanced APTEEN protocol [19] suggested that enhanced data can be transmitted continuously to the BS or on the occurrence of some special events. In the enhanced APTEEN, which is developed for the hybrid networks, different types of query handlings are presented. It is found that network lifetime of WSN is increased at the cost of increased delay by this protocol which can be enhanced by in cooperating new routing mechanisms with less complexity.

Main objective of Energy-Aware Query-based Routing Protocol [20] is to tradeoff between the energy aware and energy balancing algorithms. Its architecture consists of Zone Estimation Unit (ZEU), a Probability Computation Unit (PCU), a Data Path Selection Unit (DPSU), and an Error Detection Unit (EDU). Specific Zone (SZ), Non-specific Location (NL), Specific Location (SL) and Non-specific Zone (NZ) are four application-level API modes. Query zone is estimated by the ZEU unit which is based on low power GPS to save energy. Learning automata technique is used in the design of DPSU and PCU. The source node to transmit the query that is via SZ and SL modes will use ZONAL broadcasting. The nodes in the query zone distribute the query by adding their own position, energy level and distance to the BS. However, the delay parameter is not included in these protocols that are important parameter of the real time applications.

3.2.3   QoS-Based Routing

The major stringent requirement of the WSN is to improve the QoS metrics where quality of data and energy conservation needs to be considered. QoS metrics include energy, delay, bandwidth, time etc. that should be satisfied by the routing protocols when data is transmitted to the BS.

The first QoS based routing protocol was proposed in 2000 [21]. It used three parameters to make decision for the routing path and they are: Priority level of each packet, QoS on each path and energy resources. If any node failure occurs which cause to the change in topology then path is recomputed. It shows its validation in terms of energy consumption and fault tolerance but the overheads of maintain tables for the node and to maintain the state of the node is high and it becomes a complex problem when number of nodes is increased. An Energy-Aware QoS Routing Protocol for Wireless Sensor Networks [22] proposed routing on the basis of Qos metrics, which provides best use of the bandwidth and avoids the delay on each routing path. End to end delay, which takes into account only real time data, is considered in this work. Other factors are used to calculate the communication cost, energy stock, energy consumption rate, relay enabling rate, sensing state cost, maximum connections per node and error rate respectively.

This protocol has shown the performance for the real time and non real time traffic but as the number of nodes is increased, it becomes difficult to cope up with the quality for the real time traffic and buffer size of the nodes is always limited which is used in the queuing model. By increasing the size of the buffer it is found that more distance of the gateway from the nodes will increase and hence more delay is introduced.

Residual energy, Available buffer size, and Signal-to-Noise Ratio (SNR) parameters are used to predict the next hop through the path construction phase [23]. From these required paths, few paths are selected for the real time data and few for no real time traffic. Data transfer is accomplished in two phases as (1) Packet Segmentation and encoding; (2) Data forwarding and Recovery. Reassembling of the messages is not possible until all the segments are arrived at the destination so this phenomenon does not suit the real working of the WSN. A routing protocol for periodic and event-based data reporting with congestion control mechanism is proposed in QoS routing approach [24]. Data packets are classified in to event-based reports with higher priority and periodic reports with lower priority. The objective of this routing is to find the next hop or the node, which have high residual energy, high load and high link quality. The process of barrier mechanism is divided into the phases of the ring/barrier formation, selection of representative(s), data aggregation and forwarding mechanism, ring/barrier repair, ring/barrier enlargement/shrink and ring/barrier termination. It outperforms over SPEED protocol in delay and transmitted number of packets but it works well for little number of nodes. With large number of nodes and packets the packets drop ratio will increase and nodes which act as the barrier and do not detect the events shows the wastage of the nodes, rather they can be used in network processing tasks. The Probability of Packet Sending (PPS), Average Probability of Packet Receiving (APPR) and Interference of link between two nodes; parameters are used to find out the suitable link with QoS metrics in QEMPAR [25]. In energy consumption and end-to-end delay it shows improvement over the other protocols but complexity of this protocol is also high; which adds the extra processing at the node level.

Neighbor selection algorithm combined with geographic routing mechanism to provide multi-objective QoS routing for different application requirements is suggested in MQoSR [26]. Multiple QoS classes are formed according to the requirements of the application on the basis of energy, reliability, and delay. Reliability is the major factor in the QoS metrics that is ratio of number of packets received to the number of packets generated. In multi QoS optimization routing problem, the multi disjoint paths are find by the minimum energy consumption, minimum delay and maximum reliability functions. It uses the node-disjoint paths between the source and the sink from the n nodes. It outperforms over other protocols in terms of routing overhead, end to end delay, data delivery and energy consumption. Decision on the basis of different costs and then sorting of the paths for the final decision increases the energy consumption and can be implemented only on the small scale network. A trustworthy scheme with other QoS metrics, which considers security as an essential aspect of the routing is proposed in Trust-Aware Secure Routing Framework protocol [27]. After reception of the packet the nodes check whether the evaluated node is its neighbor, if not then it does not respond. A direct and indirect trust path is evaluated through other nodes. If the intermediate nodes have the optimal path from them then they will reply back with their trust factors. If any node does not participate or help in routing to save its energy then its trust factor is decreased. If in future the intermediate node of the optimal path could not forward the packet then it will transmit that updated information to the source node that will try to find out the new optimal path. Their neighbors observe the behavior of the nodes. It eliminates the effects of bad mouth attacks and collusion attacks.

Undoubtedly this protocol computes the trust factor of the neighbor nodes but hidden node problem cannot be solved by this algorithm and not discussed here. Including the expose of the hidden terminals security problem can optimize this algorithm.

3.2.4   Negotiation Based Protocols

The main aim of this category (Fig. 3.4) of protocols is to distribute information to all the nodes in efficient manner. Two protocols flooding and gossiping are used for this purpose. In flooding, the information is required make to available at each node by redundancy and in gossiping the information is transmitted by the node having information to its neighbor node to avoid the redundancy. Topology is not maintained in both the categories, which make them simple, but both the categories crash when the size of the network becomes large because of wastage of resources and delay. Multiple copies of same data is transferred for many times which gives rise to rise traffic implosion affecting the lifetime of WSN. To overcome these problems resource aware and negotiation based protocols are proposed by SPIN protocols family. Redundancy of the data in SPIN is avoided by the use of metadata, a high level descriptor of data, as a negotiation [28]. A node, which is interested in data, can forward its request to the node with that content. This facilitates unnecessary data

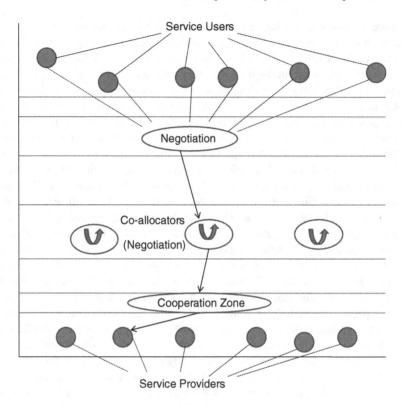

Fig. 3.4 Negotiation routing

will not flow in the network and will avoid the traffic implosion. Implosion and overlap problems are solved by SPIN-PP. Resource awareness is possible by SPIN-BC. Recovery from the packet losses is proposed in SPIN-RL. These protocols have shown the improvement over the flooding in energy efficiency and the network lifetime. Routing decisions are dependent on the specific applications having the knowledge of topology, data layout and state of resources. Format of the metadata is specific to the application. Applications of SPIN calculate the routing cost in terms of energy, processing of data, transferring and reception of data. It is three-way handshake protocol. Pros of this protocol are that own data can also be aggregated with source node data and sent with the same only. In SPIN-BC the advertised message is received not only by the destination node but by the nodes in the range. Requested data is transmitted once only and information of the sender is sent with the data. In SPIN-RL the node requests the data from its neighbors, if it is not responded back with in specific time interval then it requests the data to the list of neighbors by marking them as the destination. SPIN-PP and SPIN-EC assumed to have unlimited amount of energy and this is not possible in the wireless network and energy is limited, so we need a way to formalize the solution to conserve energy with limited energy. A uniform format of the data message can work well with the applications in communication rather than having the different formats. Maintenance of the information about neighbor is not easy task, which is required in SPIN protocols because of the failure of the nodes after sometime. To wait before the transmission of data introduces unwanted delay in network, which is not suitable for real time applications. And above all negotiation at local level cannot cover the entire network so this protocol cannot be used for the critical event detection applications. The Main aim of Interference-aware multipath routing in wireless mesh network [29] is to improve end-to-end throughput. The links, which have low channel utilization, are removed from the topology. Greedy heuristic based multipath routing algorithm is used with control of topology. The link capacity is updated every time whenever the node and its neighbors forward the aggregated data. Every node has the information about the coordinates and it is centralized algorithm. Two links are said to be interfering with each other when they are in each other range otherwise they are stated as non-coordinated. All coordinated links share the channel capacity with the link. The constraint of coordinated interference ensures that sum of flow demands of all coordinated links must be less than or equal to the available effective capacity of link. The coordinated links are represented by *Lco* and non-coordinated by *Lnco*.

During the routing process the dependencies of non-coordinated links are removed and it is linear programming model. But by eliminating the links, some paths are eliminated from the topology and network cannot be termed as fully connected and to transmit the data smoothly and without redundancy, it is required that network is connected.

3.2.5   Coherent Based Protocols

Energy efficiency routing algorithm employs various techniques to conserve energy, as transmission of required data only, reduction in redundant data etc. Routing methodologies also take care that routes should not be congested with heavy load of data and there should be alternate paths to transmit the data from different paths in case of congestion. Some techniques allow the preprocessing of raw data on the nodes before transmitting it (non-coherent) and some allow only a very little amount of processing at the source end (coherent) and data is processed at the aggregated site. Three phases for the Non-coherent algorithm are Target Detection, Data Collection, and Pre-processing; Membership Declaration and Central Node Election [30]. In the first phase the target is identified and data is gathered about it. Preprocessing is accomplished at the local level. Decisions of the local level can be overridden by the sink node, but the node should take part in the cooperative function or not, becomes clear with it. If a node is capable of participating in the cooperative routing then it enters into the second phase and declare itself the member of the current topology. Afterwards the member nodes compete for the central node election. All the nodes transmit their information about their energy and processing power. This phase is categorized into two parts that is (1) Single Winner algorithm (SWE), (2) Spanning Tree algorithm (ST). All the nodes exchange information in the former (SWE) and spanning tree is formed in the later (ST) process. In this algorithm, each node has best n nodes' data rather than only one best node (SWE). Coherent is more complex than non-coherent in terms of scalability, control overhead and delay. Two scenarios are considered in direct localization for sources in the near field and direction of arrival estimation for sources in the far field [31]. There are many conditions to be considered in the coherent array techniques like, equalization of mismatched responses under blind conditions and known conditions, partially calibrated array etc. A suitable mixed approach with beam forming and synchronization is required to best utilize the coherent techniques. In energy efficient coordination algorithm for topology maintenance [32] local decisions are made by the nodes about their participation in cooperative routing. Number of coordinators (CNs) is enough so that each radio is in the range of them. Packets destined for the sleeping nodes are buffered to avoid loss of data and made available to the node as soon it gets awakened. If all the nodes have same energy, then to decide the k as the coordinator node depends upon its neighbor connectivity provided to it from its all neighbor nodes. Nodes with higher value of neighbor connectivity are elected as the coordinators because more nodes will be connected with them. In case of unequal level of energy, the residual energy (Er) of the node is compared with maximum energy (Em) of the node to be elected as the CN and then the back off delay is computed.

SPAN improved the network lifetime at the cost of increased amount of delay. It is suitable for the small-scale applications. Its performance will be declined it the network size is increased. And comparison of the energy at the local level also introduce extra processing which needs to be solved by some other novel approach. Some other protocols DD, SPIN etc. described in the previous section fall into same category and same improvement as described there is needed in these routing protocols.

3.3 Comparison

Multi-hop routing protocols, based on network structure and reliability, focus on the various aspects like reliability, time, energy etc. parameters. Multipath protocols have high scalability and high reliability. Transmission of redundant data is required in many applications but it may have adverse affects on network lifetime and throughput. Selection of multipath routing based on quality of service metrics is preferable to avoid these problems. Some multipath protocols proposed in past, are based on these aspects like MMSPEED, MCMP etc. Coherent and negotiation based protocols generate long stream of data so energy optimization transmission become necessary for these protocols. Hybrid protocols are more advantageous for real time applications. Query based protocols are less scalable and reliable e.g. rumor routing and direct diffusion, therefore this routing can be merged with multipath and QoS routing, to exploit the benefits of redundancy, scalability and energy constraint issues. All network operation and reliability based protocols are summarized in Table 3.1 based on the different parameters.

Table 3.1 Comparison of routing protocols based network operation and reliability

Protocol	Classification	Scalability	Data aggregation	Energy efficient	Failure recovery	Network type	Load balancing	Latency	Reliability	Mobility	Location awareness
MMSPEED (2006)	Multipath and QoS	High	Yes	Good	No	Homogeneous	No	Low	Good	Stationery	Yes
Energy efficient [63] (2006)	Multipath	High	Yes	Good	Yes	Homogeneous	Yes	Average	Good	Stationery	No
MCMP(2007)	Multipath and QoS	Average	Yes	Average	Yes	Homogeneous	No	Low	Good	Stationery	No
REER (2006)	Multipath	High	Yes	Good	Yes	Homogeneous	No	High	Good	Stationery	Yes
DCHT(2010)	Multipath	Average	Yes	Average	No	Homogeneous	No	Low	Less	Quasi-stationery	No
EQMH(2012)	Multipath and QoS	High	Yes	Good	No	Homogeneous	Yes	Average	Average	Quasi-stationery	No
Beizer based [70] (2014)	Multipath	Average	Yes	Good	No	Homogeneous	Yes	Low	Average	Stationery	Yes
Directed diffusion (2003)	Query based	Low	Yes	Average	No	Homogeneous	No	Average	Less	Mobility limited	No
Rumor routing (2002)	Query based	High	Yes	Good	Yes	Homogeneous	No	High	Good	Stationery	No
COUGAR (2002)	Query based	Less	Yes	Average	No	Homogeneous	No	High	Average	Stationery	No
ZRP (1997)	Query based	Less	Less	Less	No	Homogeneous	No	High	Less	Highly mobile	No
EQR (2012)	Query based	Average	Less	Average	No	Homogeneous	No	Average	Less	Stationery	No
ERP (2014)	Query based	Average	Yes	Good	Yes	Heterogeneous	No	High	Good	Mobile	BS Aware of location of nodes

(continued)

Table 3.1 (continued)

Protocol	Classification	Scalability	Data aggregation	Energy efficient	Failure recovery	Network type	Load balancing	Latency	Reliability	Mobility	Location awareness
EQSR (2010)	Multipath and QoS	Less	Less	Average	Node failure probability 0.05	Homogeneous	Yes	High for non real time data than real time data	Average	Stationery	No
QoS based routing protocol (2010)	QoS based	Average	Less	Average	Yes	Homogeneous	No	Event based	Good	Stationery	Yes
QEMPAR (2011)	QoS based	Less	Yes	Less	Yes	Homogeneous	No	Average	Good	Stationery	Yes
Demodulation and forward (2012)	QoS based	Less	Yes	Less	No	Homogeneous	No	High	Good	Stationery	Yes
MQoSR (2013)	QoS based	Less	Yes	Average	Yes	Homogeneous	No	Average	Good	Stationery	No
TSRF (2014)	QoS based	Less	Less	Average	Yes	Homogeneous		Less	Good	Stationery	No
SPIN (2002)	Negotiation based	Average	Yes	Average	No	Homogeneous	No	Average	Average	Mobile nodes	No
SAR (2001)	Coherent based	Less	Yes	Less	No	Homogeneous	No	High	Less	Stationery	No
SPAN (2002)	Location and coherent based	Less	No	Average	No	Homogeneous	No	Average	Less	Quasi-stationery	Yes

3.4 Summary and Future Trends

All multi-hop routing protocols have common aim and try to maximize the network lifetime. Overall network techniques under network operation and reliability are, coherent based, negotiation based, QoS based, Query based and Multi path based routings. WSN in future require embedding many devices to interact with and monitor physical world. Nodes cooperate with each other to perform high-level tasks. Although widespread efforts have been put forth so far but still there are some challenges that require extensive solutions of routing issues. Sensors are tightly coupled with real world and are featured by small footprints therefore future research would need to address the various issues of quality of service, energy, and time.

References

1. Son D, Krishnamachari B, Heidemann J (2006) Experimental study of concurrent transmission in wireless sensor networks. In: Proceedings of the 4th international conference on embedded networked sensor systems (SenSys'06), Boulder, CO, USA, 31 Oct–3 Nov 2006, pp 237–250
2. Kang J, Zhang Y, Nath B (2004) End-to-end channel capacity measurement for congestion control in sensor networks. In: Proceedings of the 2nd international workshop on sensor and actor network protocols and applications (SANPA'04), Boston, MA, USA, 22 Aug 2004
3. Ganesan D et al (2001) Highly-resilient, energy-efficient multipath routing in wireless sensor networks. ACM SIGMOBILE Mob Comput Commun Rev 5(4):11–25
4. Felemban E, Lee C-G, Ekici E (2006) MMSPEED: multipath multi-speed protocol for QoS guarantee of reliability and timeliness in wireless sensor networks. IEEE Trans Mob Comput 5 (6):738–754. doi:10.1109/TMC.2006.79
5. Ming Lu Y, Wong VWS (2007) An energy-efficient multipath routing protocol for wireless sensor networks. Int J Commun Syst 20(7):747–766
6. Lee M, Wong V (2005) LPT for data aggregation in wireless sensor networks. In: Proceedings of IEEE globecom, St. Louis, Missouri, Dec 2005
7. Lin L, Shroff N, Srikant R (2005) Asymptotically optimal power-aware routing for multi-hop wireless networks with renewable energy sources. In: Proceedings of IEEE infocom, Miami, Florida, Mar 2005
8. Huang X, Fang Y (2008) Multiconstrained QoS multipath routing in wireless sensor networks. Wirel Netw 14:465–478, doi:10.1007/s11276-006-0731-9
9. Chen M, Kwon T, Mao S, Yuan Y, Victor C, Leung M (2008) Reliable and energy-efficient routing protocol in dense wireless sensor networks. Int J Sen Netw 4:104–117. doi:10.1504/IJSNET.2008.019256
10. Li S, Neelisetti RK, Liu C, Lim A (2010) Efficient multi-path protocol for wireless sensor networks. Int J Wirel Mob Netw 2(1):110–130. http://airccse.org/journal/jwmn_current10.html
11. Dehnavi M, Mazaheri MR, Homayounfar B, Mazinani SM (2012) Energy efficient and QoS based multi-path hierarchical routing protocol in WSNs. Int Symp Comput Consum Control (IS3C) 414–418. doi:10.1109/IS3C.2012.111
12. Wan S (2014) Energy-efficient adaptive routing and context-aware lifetime maximization in wireless sensor networks. Int J Distrib Sens Netw 2014(321964):16 pp. doi:10.1155/2014/321964

13. Iqbal F, Javed MY, Naveed A (2014) Interference-aware multipath routing in wireless mesh network. EURASIP J Wirel Commun Netw. doi:10.1186/1687-1499-2014-140
14. Intanagonwiwat C et al (2003) Directed diffusion for wireless sensor networking. IEEE/ACM Trans Netw 11(1):2–16
15. Manjeshwar A, Agrawal DP (2001) TEEN: a routing protocol for enhanced efficiency in wireless sensor networks. In: Proceedings IPDPS'01, San Francisco, CA, Apr 2001, pp 2009–2015
16. Manjeshwar A, Agrawal DP (2001) APTEEN: a hybrid protocol for efficient routing and comprehensive information retrieval in wireless sensor networks. In: Proceedings IPDPS'01, San Francisco, CA, Apr 2001, pp 2009–2015
17. Braginsky D, Estrin D (2002) Rumor routing algorithm for sensor networks. In: Proceedings of the 1st ACM international workshop on wireless sensor networks and applications. ACM, New York
18. Yao Y, Gehrke J (2002) The cougar approach to in-network query processing in sensor networks. ACM Sigmod Rec 31(3):9–18
19. Manjeshwar A, Zeng Q-A, Agrawal DP (2002) An analytical model for information retrieval in wireless sensor networks using enhanced APTEEN protocol. IEEE Trans Parallel Distrib Syst 13(12):1290–1302
20. Ahvar E, Serral-Gracià R et al (2012) EQR: a new energy-aware query-based routing protocol for wireless sensor networks. Wired/Wirel Internet Commun 7277:102–113 (ISBN: 978-3-642-30629-7)
21. Chen S, Nahrstedt K, Shavitt Y (2000) A QoS-aware multicast routing protocol. IEEE J Selected Areas Commun 18(12):2580–2592. doi:10.1109/49.898738
22. Akkaya K, Younis M (2003) An energy-aware QoS routing protocol for wireless sensor networks. In: Proceedings of the 23rd International conference on distributed computing systems workshops, 2003. IEEE, New York
23. Ben-Othman J, Yahya B (2010) Energy efficient and QoS based routing protocol for wireless sensor networks. J Parallel Distrib Comput 70(8):849–857
24. Fonoage M, Cardei M, Ambrose A (2010) A QoS based routing protocol for wireless sensor networks. In: IEEE 29th international performance computing and communications conference (IPCCC), 2010. IEEE, New York
25. Heikalabad SR et al (2011) QEMPAR: QoS and energy aware multi-path routing algorithm for real-time applications in wireless sensor networks. arXiv preprint arXiv:1104.1031
26. Alwan H, Agarwal A (2013) Multi-objective QoS routing for wireless sensor networks. In: International conference on computing, networking and communications (ICNC), 2013. IEEE, New York
27. Duan J et al (2014) TSRF: a trust-aware secure routing framework in wireless sensor networks. Int J Distrib Sens Netw 2014
28. Kulik J, Heinzelman W, Balakrishnan H (2002) Negotiation-based protocols for disseminating information in wireless sensor networks. Wirel Netw 8(2/3):169–185
29. Iqbal F, Javed MY, Naveed A (2014) Interference-aware multipath routing in wireless mesh network. EURASIP J Wirel Commun Netw 2014(1):140
30. Sohrabi K et al (2000) Protocols for self-organization of a wireless sensor network. IEEE Pers Commun 7(5):16–27
31. Chen JC et al (2003) Coherent acoustic array processing and localization on wireless sensor networks. Proc IEEE 91(8):1154–1162
32. Chen B et al (2002) Span: an energy-efficient coordination algorithm for topology maintenance in ad hoc wireless networks. Wirel Netw 8(5):481–494

Chapter 4
Multi-hop Network Structure Routing Protocols

Abstract With other quality of service metrics, scalability and delay metrics are major design attributes for WSN routing protocols. A single level routing can increase the load on gateway nodes which can cause latency and reliability issues. But multilevel routing can overcome this by reducing communication distance among nodes. Distance can be calculated among the nodes via location based routing. But energy aware routing is are beneficial in WSN. With distance calculation, end –to- end delay attribute is considered as important attribute of WSN. This delay is considered in QoS metrics aware protocols. In this chapter, classification of multi hop network structure protocols is done in three categories (i) hierarchical routing protocols (ii) flat routing protocols and location based routing protocols. We have thrown light on the different parameters of comparative routing protocols with research insights.

Keywords Network structure · Routing · Flat · Location based · Hierarchical · Comparison

4.1 Introduction

At the network layer, the main issue is to invent the energy efficient and reliable routing, to transmit data from source to destination, for maximization of lifetime of the network. Major challenge in sensor network is the routing because of various features that differentiate them from wireless ad hoc networks. (1) All applications of WSN necessitate the flow of data to from multiple sources to the specific sink which is contrary to typical networks. (2) Sensed data can be redundant as many sensors can generate the same data within the neighborhood. So redundancy should be reduced by the energy efficient routing protocols. (3) Global addressing scheme is not possible due to random deployment of the sensor nodes, so typical protocols cannot be used in WSN. (4) Resource management is required for the sensor nodes in terms of on-board energy, limited processing capacity, low memory and power.

© The Author(s) 2016

S. Rani and S.H. Ahmed, *Multi-hop Routing in Wireless Sensor Networks*,
SpringerBriefs in Electrical and Computer Engineering,
DOI 10.1007/978-981-287-730-7_4

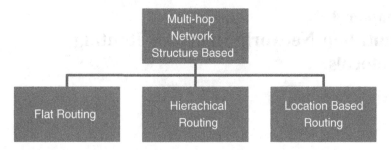

Fig. 4.1 Multi-hop routing protocols based on network structure

Many new routing algorithms have been proposed due to above-mentioned differences. These routing algorithms are either application specific or according to the
architecture requirement, which also considers the features of sensor nodes. These
routing protocols can be classified as location based, hierarchical based or data
centric (Fig. 4.1). Objective of the hierarchical protocols is to form the clusters of
the nodes so that cluster heads could aggregate the data and remove redundancy of
data to save energy. Location based protocols use the position of the nodes to
transmit data to the specific location (uni-casting) rather than to whole the network
(broadcasting of data). On the basis of network the network can be categorized into
flat, hierarchical and location based routing protocols. In flat network all nodes have
equal responsibilities and in hierarchical network some nodes are superior to others
as they play the role of CHs or relay nodes. In location based protocols data is
transmitted by exploiting the position of the nodes. In flat routing protocols, the
routing can be of reactive, proactive or hybrid type, which can be further, divided
into table driven and demand driven. In proactive the information about the routes
is stored in the table and is known in advance. In reactive protocols the routes are
computed when the nodes claim them. Hybrid protocols use both the techniques.
Some protocols are also known as cooperative protocols where data is aggregated at
one node for further processing and afterwards it is forwarded to the BS.

4.2 Network Structure Based Routing Protocols

There are some issues related to the network structure routing protocols, which are
discussed as below:

(a) Application Specific: Network design is based on the requirement of the
    applications. Some applications need minimum overhead for sensor node's
    communication and need minimum time but some application need reliability
    and energy conservation, so flat and hierarchical routing is preferred respec-
    tively for both.

(b) Global Identity: It is not possible to assign unique global identifier to each node because of the large number of nodes in the network. So data gathering is accomplished by data centric routing and flooding is used to broadcast the query of the source node. So flat routing is used for such types of networks.

(c) Location Awareness: Some applications require the location of the nodes to be known as data gathering is based on the locations. So position of the nodes is detected by using the GPS (Global positioning System) antennas but they consume more energy for this purpose. Other routing techniques are required to be developed for the awareness of the locations of the sensor nodes.

(d) Load Balancing: To maximize the network lifetime, load balancing is the god feature for the WSN. To organize the nodes into clusters is the best solution to gain this objective.

(e) Network Partitioning: Nodes, which are near to the base station, they deplete their energy faster than the other nodes and which can create the hop spot problem. Hot spot problem arises when some nodes are unable to communicate with the BS due to the failure of some nodes, which caused the network partitioning. This problem can be tackled by the use of hierarchical networking.

4.2.1   Flat Structure Routing Protocols

As each node play the same role in this type of network, so data is transmitted to the BS by using the neighbor node (Fig. 4.2) as the relay node. SPIN, Directed Diffusion, Rumor Routing etc. fall into this category which is discussed above. Some other protocols are discussed below.

1. A Highly Adaptive Distributed Routing Algorithm for Mobile Wireless Networks is developed for the highly dense and mobile networks [1]. Nodes maintain the information of their neighbor nodes only. It is source initiated and it proposes the loop free routes. Routes are created only on demand and if any change in topology requires the new paths, then new routes are established to transmit the data. Full reversal and partial reversal methods are used to maintain the links. If any node does not have any outgoing link then direction of its entire incoming links is reversed. Link reversal process is not suitable for the routes with highest traffic. Maintenance of the neighbor nodes and its' links give more burden on the tiny nodes and it adds more energy expenditure. A proactive routing protocol for dense network selects the multipoint relay of the nodes, which broadcasts the message for that node [2]. Nodes do not have to maintain the information about all the nodes but only about multipoint relays (MPRs). Each node sends its control packet periodically for the awareness of the packets, which can be lost due to collisions. But this protocol can be used only in the applications where all the information can be maintained in the tables and not suitable for the low memory nodes. TBRPF is used for point-to-point and broadcasted links [3]. It shows the improvement over flooding b considering the

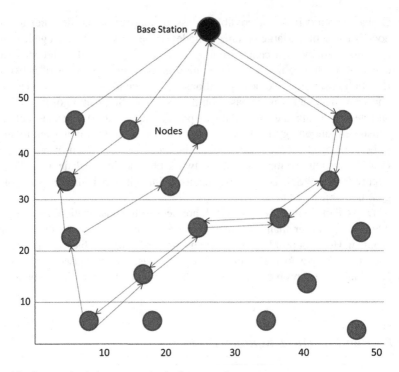

Fig. 4.2 Communication among nodes in flat network

fact that leave nodes in the trees do not require forwarded control messages. With this minimum hop forwarding method it does not ensure the correctness of the algorithm. And It also increases the computation cost by extra processing of the nodes. Minimum hop routing does not ensure the minimum consumption of the energy.

In Fish Eye State Routing [4] nodes exchange the information about the links with each other and it depends upon the distance to destination. In real time applications the delay is not tolerable which is introduced by this exchange. To examine the scope levels and radius size at the time of transmission is not feasible solution for the scalable applications. Difference between the neighbor and the gradient is computed to route the packet the on link with large gradient to ensure the energy efficiency [5]. But it does not make sure how the routes are changed if the node with DCE fails. A Scalable Solution discovers and updates the information of the minimal path with one message only. The message with dynamic cost information is forwarded with the minimum cost path [6]. It comes up with the idea that a node should defer broadcasting for some time interval which is equivalent to optimal cost of the node. But the information is not acceptable in some applications e.g. event monitoring, security surveillance etc. In another scalable algorithm the querying node requests the data from the optimal sensors by using the information utility measures as Mahalanobis distance measure [7]. In the other case the sensor node updates the belief factor and sends the information to the neighbor node for the best estimation. If the data is not reachable to the destined node with in specific time interval that issue is yet to be resolved. A Novel Gradient Approach for Efficient Data Dissemination in Wireless Sensor Networks [8] is divided into three phases gradient set up phase, data dissemination phase, and gradient reconfiguration phase. In the gradient set up phase the sink transmits the message for the hop count process to start to the nearest neighbors and it sets the value zero and the those nodes responds with the hop count plus one as their initial value is set to null. This process continues as a neighbor node transmits this hop count value of one to other nodes and they set their value to two and so on. This protocol is simple and valid for the end to end reliability. But energy conservation is not possible by this protocol. No doubt there is tradeoff between the reliability and the energy parameter but to restart the process of gradient again and again is not feasible on the limited battery power sensor nodes. In the absence of fault tolerance no protocol could give the optimal results.

The distances between the nodes are assumed uniform in both ways vertically or horizontally and diagonally in a flat protocol [9]. But the distance between the nodes cannot be equal in most of the applications e.g. nodes are deployed randomly in the environment applications. So this protocol will not fit well to the applications with mobile nodes or semi static nodes. As all metrics have their own pros and cons. It is not suitable for the query and event driven routing.

4.2.2 Location Based Protocols

Location based protocols are based on the position of the nodes to search the path among the nodes or to transmit the data to the specific locations rather than to the whole network. The location of the nodes can be finding out by the received signal strength of the nodes or by using the nodes with GPS antennas. The sensor nodes can also share the information about their locations by the exchange of the messages. The energy of the sensor nodes can be conserved by these protocols as data is transmitted to specific nodes only. Other nodes can go into sleep mode at that time. Battery power of the nodes will be less consumed in searching the route for the transmission of data due to the location aware nodes. In this section we have discussed some location aware protocols.

Energy conservation is achieved in Geography-informed routing method by using application and system information to turn off the radios for long period of time. The nodes except the source and the sink go to the sleep and wake mode alternatively. All the nodes in the adjacent grids can communicate with each other. The distance r, between the two grids should not be greater than nominal radio range R. The largest distance between farthest two nodes in adjacent grid is computed two check the virtual grid. A node can be in three states sleeping, discovery and active. In the discovery state the control messages of the nodes are exchanged within the same grid and this message is the tuple of grid id, node id, node state and estimated node active time. To tackle the mobility problem of the active nodes in the grid the sleeping interval of the nodes is kept short than the expected time interval leaving the grid (engt) of active node. *Engt* is based on the speed *s* and the size of the grid size *r* that is r/s.

This protocol behaves like other ad hoc network protocols but it conserves more energy by turning off the radios. By conserving the energy, the delay factor has been increased and the reliability has been decreased as nodes turned off their radios repeatedly and then they compute the value for the virtual grid. Exchange of the control messages after the nodes wake up also consumes energy and it is difficult to predict the nodes mobility with accuracy to know its grid value. Main objectives of loop free routing techniques are position based routing, minimum hop count and communication overhead, single path routing, scalability and guaranteed message delivery [10]. It has shown that the flooding methods f-GEDIR, f-MFR and hybrid single path/flooding are better than LAR2. It has used some localization techniques to find out shortest paths, small flooding ratios, and excellent delivery rates. But battery power, which is the major constraint of the WSN, is not considered here. Minimum hop count is not the good parameter to find out the end-to-end delay. Percolation theory is used to recognize the critical network density range, which has combined the greedy, and the face routing [11]. It is not possible to have the knowledge of the exact location of the mobile nodes and minimum hop metric is not suitable in the real network as the shortest distance computed with minimum hop metric can be larger than the distance computed with Euclidian distance metric. So energy computation by this method may not give the correct measure of the energy expenditure. Non-deterministic routing is opted in state free robust protocol in which next hop candidates compete with each other for the forwarding of the packets [12]. This process eliminates the need of the maintenance for table of neighbor's information. It promises to give the solution of close to 100 % packets transmission. But hearing the communication of the neighboring nodes is wastage of energy. To discover the forwarding nodes towards the destination for energy efficient shortest path consumes more energy than transmitting packet on the pre-defined path. Main objectives of SELAR protocol are scalability, energy efficiency and fault tolerance [13]. It uses the location of the nodes to forward the packets and assumes the sensor nodes with GPS antennas. Each node maintains the state table with attributes energy, timestamp, location and neighbor ID. The nodes within the radios range of the nodes are known as the neighbor node. It has shown its improvement over LEACH and MTE. But it does not present the feasible solution for the application where BS cannot be moved. It says the BS will start its movement after the nodes nearest to the BS lost their energy. And robot will be used to move the BS that seems virtual and not realistic solution of the WSN problems. In the geographic routing convex hulls decide the most likely direction towards the destination [14]. But trees with the highest depth cannot accurately approximate the voids, as there is large number of hops between the roots to leaf. So it is not well suited for the dense networks. EAGRP is similar in working like other geographical protocols that the nodes are aware of the neighboring node's energy and their location but with difference that source node calculates its distance from the destination and its neighbors distance from destination [15]. Calculation of the distance from each node to the destination at every transmission is energy and time consuming and not valid for the dense networks. Instead of using flooding algorithm location based routing algorithm uses the location information to confine the flooding route searching space to a smaller estimated cylindrical zone [16]. By using this strategy it automatically adjusts the radius of the cylindrical zone based on Bayes' theorem. The route request discovery is done in the smaller area rather than the whole area. The performance of protocol depends upon the RZone (Smallest zone) selection. If it does not selected carefully then it behaves like flooding. Number of transmissions in link aware protocol (ETx) is based on the state of the link. But the metric ETx has been modified in recent routing protocol to optimize the geographic algorithm for vehicular ad hoc networks [17].

Delivery probability is measured by window lasting for w seconds. Sender can compute this probability at any time t for the broadcasting for average period. But ETx does not take into account the mobility factor that is considered in this protocol. Hello packets are broadcasted rather than the dedicated probe to measure the delivery probability.

In this way, the node selects the bets neighbor node on the basis of the link state to avoid the retransmissions. This scheme is used to enhance the packet rate but is does not consider the delay and the energy metrics that can be increased for this processing. Moving vehicles change their locations frequently and so as information is updated in the sensor nodes. But location of the information must be predicted so as the sensor nodes may have the timely information about the location [18]. This protocol predicts the location that depends upon the speed and the heading direction of the vehicle. It is divided into three components location predictor, packet forwarding, and buffer management. The location predictor predicts the location on the basis of the location.

In the next phase, the nodes with high priority will forward the packets first. The nodes, which are close to the destination, they are given more weight that is they are given more priority than others are. In the buffer management when the buffer is full then the nodes with highest residual distance and low residual energy are replaced first. But by replacing the packets for the large distance it will encounter delay for those replaced packets and the packets that should be deliverable in time, will not reach to the destination.

4.2.3   Hierarchical Based Protocol

The sensor nodes in the network can be organized into various clusters, which are managed by different cluster heads (Fig. 4.3). This type of organization may be known as hierarchical or the layered design in which data may be passed from one cluster head to another before reaching the sink. Cluster heads supposed to be more powerful where they require processing the aggregated data otherwise their energy level may be equal to other nodes. The nodes with lower energy level may perform the task of sensing the data whereas the nodes with higher energy level may

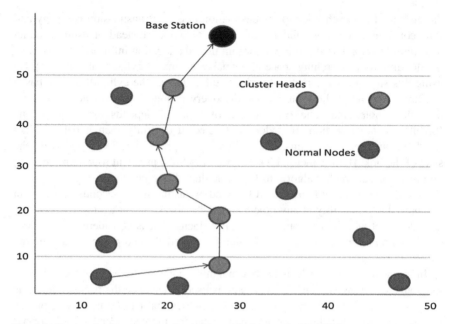

Fig. 4.3 Communication among nodes in hierarchical network (Via CHs)

function as the cluster heads. Major issues in clustering topology is how to from the clusters and how to elect the cluster heads. Clustering can be single hop or multi-hop depending upon the communication of the CHs. In single hop communication data is forwarded to the BS directly by the CHs and in the multi-hop data is forwarded from the long distant (from BS) CH to the short distant (from BS) CH. Main objectives of the clustering are load balancing, enlarged connectivity, fault tolerance, long network lifetime etc. Cluster heads can be mobile or stationery depending upon the requirement of the application. Various cluster-based protocols have been proposed in recent past and few of them are discussed here.

A networking environment in which users are mobile and interference occurs because of the multiple communications, the traffic must be controlled [19]. Adaptive clustering is proposed for this type of the environment in which multimedia and data traffic both are supported. Transmitter based coding is assigned to the nodes to avoid the inter-cluster and intra cluster interference. It has shown the performance equal to TDMA cluster scheme but with less complexity. The main objective of load balanced scheme, is to form the cluster around the high-energy gateway nodes and to balance the load among the clusters and hence to increase the lifetime of WSN [20]. According to the received information, clustering phase is accomplished on the basis of the current load of the cluster and the communication cost.

First order radio model [LEACH] is used to compute the energy expenditure. The gateway nodes act as the relay nodes through which the data of all the nodes is transmitted. Fault tolerance in case of failure of gateway node is not solved here. The collection of the information of all the nodes on the gateways is not the feasible idea in high dense network. Genetic algorithms are also proposed in literature for the cluster based applications [21]. In the agriculture field accurate uniformity is required so that sub-areas could not be overlapped with each other so mean relative derivation (MRD) is used. MRD is derived by computing deviation of the spatial density of measurements in each sub-area from the total spatial density of measurements in the entire area. Second parameter, which is application specific, is spatial density error (SDE). For network connectivity fitness function Sensors-per-Cluster head Error (SCE) is used to assure that each CH do not have more load than predefined. Sensors-Out-of-Range Error (SORE) is used for the communication of each sensor with its cluster head. Genetic algorithm decided which nodes will be inactive and which will be active. The problem with genetic algorithms is that they consume more time and using the hybrid techniques than by using only mutation and crossover functions can increase the energy efficiency. According to this clustering each node implements DWEHC individually [22]. Clustering process completes in seven iterations. For inter-cluster communication nodes contend for the channel using IEEE 802.11 model to transmit the data. It has shown its improvement over HEED-AMRP algorithm in terms of energy efficiency by using the balanced cluster approach. In this location based attribute work hierarchies are established in the network so that information can be transmitted at the end where it is required [23]. Various levels of clusters interact with each other to route the information and the nodes elect CH at each cluster. Levels of hierarchy are added and removed dynamically based on the queries and so as the role of CH is also changed with these levels. As compared to pure flooding systems this scheme is more robust under heavy network traffic. All the nodes within one cluster may have similar data and sink does not require the same data to be received from that cluster [24]. This redundancy in the data can be reduced by using distributed source coding scheme (DSC).

It depends upon the number of the sensor nodes N in the network and degree of correlation C. Low correlation requires small cluster sizes and high correlation requires large cluster sizes. Routing proposed in this method allows maximum data fusion at each node, which is not possible due to low memory and small battery size.

Perspective of MISO based routing system is not only the network layer but the all the three layers from physical to network layer [25]. At the physical layer energy consumption model is analyzed for the one hop and hop-to-hop communication. Then it proves that energy efficient routing with optimal cooperation set selection problem (E2ROCSS) is the NP hard problem. And it has proposed routing algorithm with minimum communication overhead to overcome this problem. In this paper we are discussing about the routing algorithms so we will not discuss about the physical layer.

It has proved its validity over SISO based routing work for WSN. To reduce the active set for the sensing task it is assumed that sensor nodes have localization capabilities [26]. Each node needs to determine the overlap of its sensing area with the sensing areas of its neighboring nodes. The total energy determines the minimum weight coverage cost and the cost of finding the location of overlap area. The weighted average of the total energies of all points that are covered by the sensing area of node, is find out as for the critically covered location (x, y) within the sensing range of the node.

In this protocol the CHs are elected in the first phase and if any CH is dead then new CH broadcasts the message of announcing itself as the CH within its communication range then active nodes which cover the entire network transmits the data with the help of CHs. Its strength lies in the optimization of coverage aware routing by minimizing the use of low energy sensor nodes in sparsely covered areas. EEHCA avoids the frequent election of the same node as CH and also introduced the concept of back up cluster head [27]. The optimum number of CHs is elected on the basis of communication radius of the clusters and overlapping with other clusters. The energy consumption is calculated in most cluster based protocols by the equations used in first order radio model as discussed in LEACH but in this protocol it has been stated that actual cluster area is circulatory so the energy consumption is to be calculated for the cluster area.

EEICCP is the hierarchical multi-hop homogeneous protocol, which has optimized the energy efficiency for the long network lifetime of WSN [28]. Its main objective is to reduce the distance covered by the nodes. Different levels of clusters are formed in which data is transmitted through the cluster coordinators (CCOs). Cluster heads aggregate the data of the local nodes and communication between clusters is made possible through CCOs. Considering the fact that local communication of the nodes with CH is direct and distance covered by the nodes at local level is greater as dimension of one cluster is assumed 200 m by 20 m can enhance it. It can be optimized in terms of time, reliability and energy efficiency. Energy aware distributed clustering (EADC) is the non- uniform nodes distribution protocol [29]. But in this protocol energy consumed in transmitting redundant data is not considered which is analyzed in SA-EADC [30]. It is based on the approach if the sensing disks of two nodes are intersecting with each other and one node is present in the communication range of other nodes then that node is redundant and can be inactive. All the active nodes select the CH for their clusters according to the received signal strength. And data is transmitted through the active nodes. It validates itself in the parameter of network lifetime.

4.3 Comparison

Table 4.1, provides the comparison of all the protocols based on various parameters. Wireless sensor networks are application specific, so we have compared the protocols based on their use in applications and these parameters are scalability, reliability, energy efficiency, fault tolerance etc. Sensor nodes can be classified as homogenous or heterogeneous on the basis of communication and transmission capabilities. The nodes, which are heterogeneous, can regulate themselves for data transmission to conserve energy. Flat routing protocols have more network throughput than hierarchical protocols but at the cost of energy. Location based protocols can route data easily because data transmission is based on position of the nodes. Nevertheless, to searching of locations repeatedly can cause the depletion of energy by nodes and hence affect the network lifetime. Hierarchical protocols are

Table 4.1 Comparison of multi-hop routing protocols based on network structure

Protocol	Classification	Scalability	Data aggregation	Energy efficient	Failure recovery	Network type	Load balancing	Latency	Reliability	Mobility	Location awareness
Gradient approach (2004)	Flat structure	Less	Yes	Average	Yes	Homogeneous	No	Depends upon the number of the nodes	Less	Semi-stationery	No
OLSR (2001)	Flat structure	High	Yes	Less	Yes	Homogeneous	No	Less	Less	Mobile nodes	No
TORA (1997)	Flat structure	High	Yes	Less	Yes	Homogeneous	No	Less	Average	Mobile nodes	No
GAF (2001)	Location based	Average	No	Average	Yes	Homogeneous	Yes	Average	Good	Mobile nodes	Yes
GOAFR (2003)	Location based	High	No	Less	No	Homogeneous	No	Less	Less	Stationery	Yes
IGF (2003)	Location based	Less	No	Average	No	Homogeneous	No	Less	Less	Mobile nodes	Yes
SELAR (2005)	Location based	Good	No	Good	Yes	Homogeneous	No	Average	Good	Sink is mobile	Yes
EAGRP (2010)	Location based	Less	No	Average	No	Homogeneous	No	Less	Average	Stationery	Yes
LSGR (2014)	Location based	Average	No	Average	No	Homogeneous	No	Less	Less	Mobile nodes	Yes
DWEHC (2005)	Hierarchical	Average	Yes	Good	No	Homogeneous	Yes	High	Average	Quasi-stationery	Yes
UCR (2007)	Hierarchical	Average	Yes	Good	No	Homogeneous	No	High	Average	Stationery	No
MISO-based routing (2009)	Hierarchical	Average	Yes	Good	No	Homogeneous	No	High	Less	Stationery	No

(continued)

Table 4.1 (continued)

Protocol	Classification	Scalability	Data aggregation	Energy efficient	Failure recovery	Network type	Load balancing	Latency	Reliability	Mobility	Location awareness
EADUC (2011)	Hierarchical	Less	Yes	Good	No	Heterogeneous	No	Average	less	Stationery	No
EEICCP (2013)	Hierarchical	High	Yes	Good	Yes	Homogeneous	No	Less	Good	Stationery	Yes
SA-EADC (2014)	Hierarchical	Average	Yes	Good	No	Heterogeneous	No	Average	Less	Stationery	No

scalable, energy efficient but can take more time to deliver the data and consequently, network throughput is decreased. The proposed schemes for small networks can all be applied to large networks due to scalability so this is the essential feature of routing protocols. While execution of routing process some processing occurs in transmission of data via transmission media that can be more efficient, if size of the data and bandwidth are unbiased.

4.4 Summary and Future Trends

Multi-hop routing protocols are classified and compared based on some parameters, which are application specific.

In these protocols, selection of CH is an important aspect and major research is how to utilize the battery of the nodes at maximum. That is why; clustering has been considered best topology for efficient routing. By keeping in view, the routing techniques of LEACH, many protocols are developed. But with clustering, multi-hop technique has revolutionized the routing and it has brought many advantages. Research can be conducted in future based on the hybrid routing techniques on many open issues. Multimedia sensor network require delay, error and jitter control. Many error-controlling schemes like forward error control and selective-repeat automatic repeat request (ARQ) can be applied to transmit data in more reliable way. Research in vehicular technology is dominating the future trends. Many government and private agencies are working on this area and in the upcoming days sensors will be deployed on the moving vehicles to capture the real time data. Security is need of all the applications, which is required with all the other features of WSN. Secure and efficient hybrid protocols necessitate constructing in near future.

References

1. Park VD, Corson MS (1997) A highly adaptive distributed routing algorithm for mobile wireless networks. In: 16th annual joint conference of the IEEE computer and communications societies, driving the information revolution (INFOCOM'97), vol 3, IEEE
2. Jacquet P et al (2001) Optimized link state routing protocol for ad hoc networks. In: Technology for the 21st century multi topic conference on IEEE INMIC 2001, IEEE
3. Bellur B, Ogier RG (1999) A reliable, efficient topology broadcast protocol for dynamic networks. In: 18th annual joint conference of the IEEE computer and communications societies (INFOCOM'99), vol 1, IEEE
4. Pei G, Gerla M, Chen TW (2000) Fisheye state routing: a routing scheme for ad hoc wireless networks. In: IEEE international conference on communications (ICC 2000), vol 1, IEEE
5. Chang RS, Kuo CJ (2006) An energy efficient routing mechanism for wireless sensor networks. In: 20th international conference on advanced information networking and applications (AINA 2006), vol 2, IEEE
6. Ye F et al (2001) A scalable solution to minimum cost forwarding in large sensor networks. In: Proceedings of the 10th international conference on computer communications and networks, IEEE

7. Chu M, Haussecker H, Zhao F (2002) Scalable information-driven sensor querying and routing for ad hoc heterogeneous sensor networks. Int J High Perform Comput Appl 16 (3):293–313

8. Han KH, Ko YB, Kim JH (2004) A novel gradient approach for efficient data dissemination in wireless sensor networks. In: IEEE 60th vehicular technology conference (VTC2004), vol 4, Fall 2004, IEEE

9. Kanavalli A et al (2009) A flat routing protocol for sensor networks. In: Proceeding of international conference on methods and models in computer science (ICM2CS 2009), IEEE

10. Stojmenovic Ivan, Lin Xu (2001) Loop-free hybrid single-path/flooding routing algorithms with guaranteed delivery for wireless networks. Parallel Distrib Syst IEEE Trans 12(10):1023–1032

11. Kuhn F, Wattenhofer R, Zollinger A (2003) Worst-case optimal and average-case efficient geometric ad-hoc routing. In: Proceedings of the 4th ACM international symposium on mobile ad hoc networking and computing, ACM

12. Son S et al (2003) IGF: a state-free robust communication protocol for wireless sensor networks. Technical report, Department of Computer Science, University of Virginia, Virginia

13. Lukachan G, Labrador MA (2004) SELAR: scalable energy-efficient location aided routing protocol for wireless sensor networks. In: 29th annual IEEE international conference on local computer networks, IEEE

14. Leong B, Liskov B, Morris R (2006) Geographic routing without planarization. In: NSDI, vol 6

15. Elrahim AGA et al (2010) An energy aware WSN geographic routing protocol. Univ J Comput Sci Eng Technol 1(2):105–111

16. Li XH, Hong SH, Fang KL (2011) Location-based self-adaptive routing algorithm for wireless sensor networks in home automation. EURASIP J Embed Syst

17. Li C et al (2014) A link state aware geographic routing protocol for vehicular ad hoc networks. EURASIP J Wirel Commun Netw 1:1–13

18. Zhu Y et al (2014) Geographic routing based on predictive locations in vehicular ad hoc networks. EURASIP J Wirel Commun Netw 1:1–9

19. Lin CR, Gerla M (1997) Adaptive clustering for mobile wireless networks. Sel Areas Commun IEEE J 15(7):1265–1275

20. Gupta G, Younis M (2003) Load-balanced clustering of wireless sensor networks. In: IEEE international conference on communications (ICC'03), vol 3, IEEE

21. Ferentinos KP, Tsiligiridis TA (2007) Adaptive design optimization of wireless sensor networks using genetic algorithms. Comput Netw 51(4):1031–1051

22. Ding P, Holliday JA, Celik A (2005) Distributed energy-efficient hierarchical clustering for wireless sensor networks. In: Distributed Computing in Sensor Systems. Springer, Berlin, pp 322–339

23. Wang K et al (2005) Attribute-based clustering for information dissemination in wireless sensor networks. In: Proceeding of 2nd annual IEEE communications society conference on sensor and ad hoc communications and networks (SECON'05), Santa Clara, CA

24. Chen H, Megerian S (2006) Cluster sizing and head selection for efficient data aggregation and routing in sensor networks. In: Wireless communications and networking conference (WCNC 2006), vol 4, IEEE

25. Zhou P et al (2009) An energy-efficient cooperative MISO-based routing protocol for wireless sensor networks. In: Wireless communications and networking conference (WCNC 2009), IEEE

26. Soro S, Heinzelman WB (2009) Cluster head election techniques for coverage preservation in wireless sensor networks. Ad Hoc Netw 7(5):955–972

27. Xin G, WH Yang, DeGang B (2008) EEHCA: an energy-efficient hierarchical clustering algorithm for wireless sensor networks. Inf Technol J 7(2):245–252

28. Rani S, Malhotra J, Talwar R (2013) EEICCP—energy efficient protocol for wireless sensor networks. Wirel Sens Netw 5:127

29. Yu J et al (2012) A cluster-based routing protocol for wireless sensor networks with nonuniform node distribution. AEU-Int J Electron Commun 66(1):54–61

30. Nokhanji N et al (2014) A scheduled activity energy aware distributed clustering algorithm for wireless sensor networks with nonuniform node distribution. Int J Distrib Sens Netw

Chapter 5
Future Research and Scope

Abstract From few years, interest in wireless sensor networks has been in potential use for many applications like border security surveillance, disaster management, field of health and objection detection in remote areas. Sensors are deployed in wide area to operate autonomously for long time in unattended environment. Sensors are equipped with low memory and limited battery power. The main job of wireless sensor nodes is to sense, gather and transmit the data to the center location. This requires the efficient routing paths to be set up between the sink and the sensor nodes which can be satisfied by various routing protocols. Industrialists, Individuals, researchers and students are trying to develop the efficient routing protocols in terms of various quality of service metrics. To develop the routing protocols it is necessary to understand the basics of WSN. In this chapter, we have given the brief overview of fundamentals of WSN, a deep insight into comparative study of various multi-hop routing protocols with future scope.

Keywords Fundamentals · Comparative analysis · Multi-hop routing · Future trends

5.1 Fundamentals of WSN

Sensor nodes are very tiny nodes, deployed randomly on geographical area. They produce a link between digital and real world by sensing and processing the real world data. Sensors help to keep away from disastrous failures, enhance productivity, save natural resources, and to develop the new applications such as smart home, intelligent medical network etc.

The sensing circuitry of nodes measures the various parameters of the environment and converts them into the electrical signal. Processing of these signals reveals certain properties about objects located to and events happening in the surrounding area of the sensor. Each sensor is equipped with embedded processors, low power radio, and limited memory. The sensors transmit the gathered data with

© The Author(s) 2016
S. Rani and S.H. Ahmed, *Multi-hop Routing in Wireless Sensor Networks*,
SpringerBriefs in Electrical and Computer Engineering,
DOI 10.1007/978-981-287-730-7_5

low power radio directly to the base station (BS) or indirectly with the data concentration center (a gateway). The low cost and decrease in the size of sensors have encouraged the practitioners to exploit the potential capabilities of sensors in a large-scale network, which can operate unattended [1–6]. A lot of survey has already been done on the routing protocols for WSN [7–14] and have discussed that unattended network sensor nodes are expected to have significant effect on the efficiency of many civil and military applications such as border security and battle field surveillance, disaster management etc. Large number of sensor nodes can be dropped by helicopter to gather data, for monitoring the area of interest. The short lifespan of the battery-operated sensors and the possibility of having damaged nodes during deployment, coverage of large area, a vast number of sensor nodes are predicted in various applications of WSN. It is visualized that sensor nodes can range from hundreds to thousands for intended applications. For example, in a disaster management a large number of nodes are deployed randomly to identify the risky areas and for rescues operations. A similar application may be the monitoring of the air pollution, where sensor nodes are deployed in various cities to monitor the concentration of dangerous gases for general public. Moreover, a WSN may be used for detection of forest fires to generate an event when a fire has started. The nodes can be equipped with sensors to control humidity, gases and temperature, which are formed by fire in the trees or vegetation.

Sensors are being widely used in the health care applications where they may offer significant cost savings and enable new functionalities that can assist the elderly people living along in the house or people with chronic diseases on the daily activities. With the help of WSN, landmines can be used for more civic use by controlling them remotely and target specific to prevent injury to animals and human beings. Applications of security in WSN include criminal and intrusion detection. However, having the limited bandwidth and power, sensor nodes is deployed typically, which increase the challenges in the management and design of the sensor networks. These design and management issues make it required making changes in all layers of the networking protocol stack. Problems regarding the physical layer and data link layer are focused on low duty cycle, energy aware MAC protocols, dynamic voltage scaling [15–19] etc. At the network layer, the main issue is to invent the energy efficient and reliable routing, to transmit data from source to destination, for maximization of lifetime of the network. Major challenge in sensor network is the routing because of various features that differentiate them from wireless ad-hoc networks. (1) All applications of WSN necessitate the flow of data to from multiple sources to the specific sink which is contrary to typical networks. (2) Sensed data can be redundant as many sensors can generate the same data within the neighborhood. So redundancy should be reduced by the energy efficient routing protocols. (3) Global addressing scheme is not possible due to random deployment of the sensor nodes, so typical protocols cannot be used in WSN. (4) Resource management is required for the sensor nodes in terms of on-board energy, limited processing capacity, low memory and power.

There are many challenges and constraints in the working mode of WSN such as unattended environment, random deployment, limited energy, low memory, limited processing capability, security, data aggregation, security etc. Therefore, many new routing algorithms have been proposed due to above-mentioned differences. These routing algorithms are either application specific or according to the architecture requirement, which also considers the features of sensor nodes. These routing protocols can be classified as location based, hierarchical based or data centric. Objective of the hierarchical protocols is to form the clusters of the nodes so that cluster heads could aggregate the data and remove redundancy of data to save energy. Location based protocols use the position of the nodes to transmit data to the specific location (uni-casting) rather than to whole the network (broadcasting of data). Data centric protocols focus on demand of the desired data or request of the data in the form of query. One more category can be considered for the routing protocols i.e. protocols, which are based on quality of service metrics (QoS). The aim of this category is to design the route of the nodes to meet the general requirements (Timely delivery, Reliability, Scalability etc.) of wireless sensor networks (WSN).In the following subsection we have analyzed the various multi hop routing protocols discussed in the preceding sections.

5.1.1 Comparative Analysis of Multi-hop Routing Protocols

Routing protocols are compared on the basis of various parameters like energy efficiency, scalability, fault tolerance, reliability, latency, data transmission; load balancing, network type, mobility, location awareness and failure recovery. Some routing protocols are scalable that is their routing approaches for the small networks can also be applied to the large networks. Some protocols like SHARP, EB-PEGASIS, AODV-DSDV, EEHCA, GBHHR etc. are scalable. Routing protocols should be adaptive in nature so that they could accommodate the change in the size of the networks. Some Genetic Protocols, which are based on the mutation and crossover, are not scalable [20] as they take a lot of time to execute if the number of nodes is increased. Next parameter in the table is data aggregation. Data aggregation is the process of collection of data by one node from other nodes. It serves many purposes. It is used to reduce the data redundancy, to enhance energy conservation and to reduce delay. When the one node aggregates data, then the redundant data is deleted before forwarding it to the other node in the route. Data can also be compressed at that node to further reduce the size of the data. As size of the data is decreased, it helps in energy conservation and also reduces delay. Many protocols use this parameter as EEUC, EEICCP, LEACH, OLSR, and TORA etc.

Due to the limited battery power of the sensor nodes; major concern of the WSN applications is to enhance the network lifetime by using novel routing approaches. After the deployment of the tiny sensor nodes, it is not possible to change or recharge the battery of the sensor nodes. So it is necessary to develop the energy efficient routing algorithms. SELAR, DODA, DWEHC, ERP, MMSPEED etc. are some good energy efficient routing protocols. Routing protocols should work even after the death of some nodes that is routing approaches should develop the algorithms in such a way that data of the nodes could be sent to the BS even if some nodes have already depleted their energy. HEED, EEHCA, MCMP, REER etc. have used failure recovery mechanism to make the protocols more reliable. According to the capabilities of sensor nodes, network can be categorized into homogeneous or heterogeneous [21]. In homogeneous network nodes are same on the basis of energy, computation and processing and their communication links are symmetric. But in heterogeneous network some nodes are more powerful than others in same parameters. In practical the assumption of homogeneous nodes is possible hardly as no two sensors can be same on the basis of physical properties. To fulfill the specific needs of WSN application sensors are required to be different in processing capabilities so heterogeneous network are more desired in real world. But for the sake of simplicity in routing algorithms and for applications where same type of nodes in terms of energy are required giving rise to the need of homogeneous network. Protocols reviewed in this paper fall into both of these categories. EEUC, SHARP, M-GEAR, LEACH-TM, PEGASIS, CCRP, MMSPEED, EEICCP etc. are homogeneous whereas LayHet, EECP, GBHHR, CHR, SEP, EADUC etc. are heterogeneous.

Nodes, which have more load than others, or the nodes, which play the role of relay nodes, deplete their energy earlier than the other nodes. Failure of some nodes can partition the network, which can abort the collection of information from some specific region. To deal with this problem it is required that load on the nodes should be balanced. Load balancing on the nodes can increase the network lifetime and network will not be obsolete. Routing protocols with load balancing algorithm more strongly assure the long network lifetime of the WSN network like HEED, SEP, EQMH, EQSR, GAF, and DWEHC etc.

Timely delivery of data is required in most of real world applications like security surveillance, health monitoring, aircraft controlling applications etc. Data should be reachable to the BS in time in real time applications. These applications cannot tolerate delay. Query driven applications also necessitate less delay. Routing strategies should take care of transmission of data so that it should be reachable in time. These techniques also take care of avoiding collisions and use CSMA/CA or CSMA/CD or TDMA techniques. CHIRON, CCBRP, DEEC, CHR, LayHet, TSRF, DCHT etc. routing protocols take less time in transmission of data.

Data can be transmitted to the BS directly by the node, which is called single hop communication. To reduce the travelling distance and to conserve the energy, data can also be transmitted to the BS via other nodes, which are known as the relay nodes. This type of communication with BS is known as multi-hop data transmission. For long distance communication, direct transmission of data is not preferable as large amount of energy is required for processing and sensing. The expenditure of energy in transmission overpowers the total energy depletion for communication and requisite of transmission power grows exponentially with the increase of transmission distance. So in terms of long network lifetime, multi hop communication is preferred over single hop communication for long distance communications. EEUC, CCRP, CHIRON, MIEEPB, DEEC, Directed Diffusion, Demodulation and forward protocol, OLSR etc. are using multihop communication.

But the protocols ZRP, ACQUIRE, APTEEN, EECP etc. are good for short distance communication rather than long distance communication as they use single hop communication. Various WSN applications have various reliability requirements. Data forwarded to the BS is loss tolerant because of redundancy of data but data coming from BS in the form of queries is required to reach the BS with more reliability. Sensed data can be reached to the BS but it should be in correct form to facilitate correct decision-making. So error control becomes the necessary condition to achieve the reliability. Reliability and fault tolerance require that more nodes should be deployed than essential nodes. But use of extra nodes can increase the more energy consumption. To increase the reliability, some routing protocols require that applications should operate in heterogeneous networks. To increase the reliability of the network some protocols also use multipath routing. If the transmission on one path is failed, other route can transmit the data. There is always tradeoff between traffic overhead and reliability and multipath routing increases the traffic overhead [22]. So, Dulman et al. has proposed the variant of multipath routing. A data packet is decomposed into K sub packets of equal size with redundancy and sent over K disjoint paths. Reconstruction of the original packet requires only small number of packets. Multi-hop routing also adds on the reliability factor in routing. SHARP, EEHCA, GBHHR, LayHet, ROUT, MMSPEED etc. are the protocols, which are good in terms of reliability because of the above-discussed features.

Based on the mobility, sensor nodes can be categorized as stationery and mobile nodes. In some applications nodes are required to be static so they do not move. But in some applications sensing is possible only by the mobility. The protocols, which consider the mobility of the nodes, they are more complex than the others, which do not. But due to the requirements of the WSN applications, this feature is necessary to include in the routing mechanisms. In some routing protocols as IEEPB, MIEEPB, IECBSN etc. sink is mobile and in others IDSQ/CADR, ZRP, ERP etc. nodes are mobile. Some applications also require that location of the sensor nodes is known. Locations can be computed by using the sensors with GPS antennas, or by received signal strength of the incoming signals or the locations can be exchanged by the nodes before communication. Localization techniques can also be used to compute the locations but they do not guarantee 100 % accuracy. On the other hand to use the GPS antenna on each node is not cost effective. But location information is useful in constructing the routing paths by computing the distances. So it is the requirement of WSN that location of the nodes is known in advance to measure the actual distance among the nodes for the forwarding of data. DCBR, TLES, PEDAP, CCRP, CHR, ROUT, MMSPEED etc. are location aware protocols. Protocols discussed here do not belong to one category only that is protocols which are discussed under the heterogeneous category they are location aware also like ROUT. If some protocols are discussed under the multipath routing they are also QoS aware protocols like MCMP. Hybrid protocols give better results than the protocols, which work on the single aspect of the WSN.

5.2   Future Scope

WSN can be implemented using any of the well-developed standards such as IEEE 802.15.4, BLUETOOTH, ZigBee, Wibree etc. [23]. The design used in Zigbee is optimized for low cost production and it provides full validation for the requirements of physical layer. However it is constrained in bandwidth and power. Bluetooth was designed to reduce the interference between wireless signals and it can handle both data and voice transmissions. IEEE 802.11 focuses on the physical and link layers. IEEE 802.11b and 802.11g standards use the 2.40 GHz band so they can interfere with cordless phones as Bluetooth and some other devices use the same frequency band. Nevertheless, 802.11a and 802.11h use 5 GHz band so it does not face interference problem by 2.4 GHz band. For ultra low power consumption, Wibree technology was developed. To transfer and receive the IPV6 packets on personal area networks (PANs), especially on IEEE802.15.4-standard networks, 6lowpan was developed which is the name of working group in the Internet area of IETF. In addition, the software also brought revolution in the field of WSN by enabling advance programming techniques for the routers, modems and sensor nodes. TinyOS is the first operating system specifically designed for wireless sensor networks. IT is written in nesC, an extension of C programming language. There are other software like SOS, MANTIS, Contiki etc. for programming in C language. Many new models have been proposed and developed to cope up with sensor nodes as Distributed Compositional Language (DCL), Protothreads, SQTL, and SNACK etc. Hardware and software developments are still need to handle the constraints and challenges of WSN. Future research will be based on the programming not only for one layer but also at more than one layer. Cross layer communication is the solution of many problems [24] and it can handle the many issues simultaneously like, frame reception probability depending upon the signal to noise ratio (SNR), reliability, fault tolerance, congestion control by splitting the data of various paths etc. Future applications require on demand services. Many applications will require large scale static or dynamic WSNs and several challenges at different levels have to be faced by it [24].

1. Sensors: Humidity, presence of objects, temperature and speed is easy to implement using tiny devices But large-scale applications require, nano sensors to be developed for all the fields to cater the need of all the applications in future.
2. Network programming: Programming of the nodes is not an easy task. In multi-hop communication, nodes need to maintain the cooperation. Cooperation of the nodes requires programming at nodes so that they could recognize their neighbors, routes could be searched easily and nodes could use the robust routing. All these requirements need advance programming techniques and platforms.
3. Security: The replication of virus on the root mote can affect neighbor nodes easily. Research on many security issues is going on but more controlled and safe communication is required in WSN. Working of WSN is different from

internet. In sensor network, a node broadcasts the message, and recipients of the nodes may store or ignore the message. Few nodes can process the message and may respond back to the sender. From response a sender can determine which nodes need maintenance. But reprogramming does not ensure the virus free atmosphere. It poses tough constraint in front of WSN. Data transmission on multiple routes can be hacked and new data can be introduced at any time. So, tight security is the demand of all the applications.

4. Power: power is the major constraint of sensor nodes. Due to depletion of energy of the nodes, nodes die unexpectedly. Routing protocols are being developed to handle this issue as more energy is consumed in data communication rather than in sensing and processing. But batteries of the nodes should be rechargeable. There should be some ways to replace the already deployed nodes. Research is already in progress in bio-chemical industry, medicine industry and environmental science, which will bring solution in the near future.

5. Privacy: In the upcoming days, sensor nodes will be less visible and will be situated at less difference. Nodes will draw energy from environment itself. Sensors are used in vehicles, human body, farms, shopping centers etc. to capture the motion, presence, and conditions of the objects. All these deployments require privacy to be maintained. A proper balance is the demand and to be maintained from the exploitation of the benefits of WSN with maintenance of privacy.

6. Node failure: Solutions to handle the unexpected node failures required to be developed so that network communication not halted by the failure of some nodes. Communication of some nodes is affected by noise or by weather conditions; solutions need to be devised to cope with these problems.

5.3   Conclusion

Wireless sensors are the requirement of many applications. SNs are used in many applications as health care monitoring systems, border surveillance tasks, environment monitoring, transportation controlling, home security systems etc. Reliability of the information is most essential module of all the applications. It depends upon the energy efficiency of the nodes, communication of the nodes, transmission type, network type, delay, security encoding schemes, fault tolerance and signal to noise ratio. From these, energy efficiency is the major concern of WSN. It depends upon the careful selection of routing and processing load of the nodes. For the real time applications, delay is not tolerable. So in these applications routing algorithms with short distance communication is required. To apply security on border, events should be carefully monitored so reliability and real time routing are compulsory features to be implemented. Network in these applications should be heterogeneous and scalable. For small-scale applications like patient monitoring, for precision agriculture etc., there is no need to implement the hierarchical clustering approach; flat protocols are more suitable than hierarchical protocols. For large-scale applications like habitat monitoring, forest monitoring, environment monitoring etc. division of network area into clusters is most suitable approach. Applications operating in dynamic environment require location aware protocols. Multipath routing can be used for failure recovery and fault tolerant purposes. Applications where loss of information cannot be tolerated like video image processing, voice processing etc. multipath routing can be used. Stringent requirements of the WSN and due to new researches; mobile nodes have become the center of attraction. These nodes collect the data autonomously and sharply from the other nodes. Speed of the mobile nodes should be carefully examined in routing algorithms. Different applications have different priorities, so routing is very critical in these applications and should be selected carefully. With hardware and software advancements, routing protocols which can increase the network lifetime and can provide reliable and timely data delivery are in demand in coming generation. So a novel method which can fulfill this objective is required ensuring the connectivity of the network under severe conditions.

References

1. Estrin D, Govidan R, Heidemann J, Kumar S (1999) Next century challenges: scalable coordination in sensor networks. In: Proceedings of the fifth annual international conference on mobile computing and networks (MobiCom '99), Seattle, Washington, August 1999
2. Katz RH, Kahn JM, Pister KSJ (1999) Mobile networking for smart dust. In: Proceedings of the 5th annual ACM/IEEE international conference on mobile computing and networking (MobiCom '99), Seattle, WA, August 1999
3. Pottie GJ, Kaiser WJ (2000) Wireless integrated network sensors. Commun ACM 43(5):51–58
4. Akyildiz IF, Su W, Cayirci E et al (2002) Wireless sensor networks: a survey. Comput Netw 38:393–422
5. Chong C-Y, Kumar SP (2003) Sensor networks: evolution, opportunities, and challenges. Proc IEEE 91(8):1247–1256
6. Wang H, Elson J, Girod L, Estrin D (2003) Target classification and localization in habitat monitoring. In: Proceedings of IEEE international conference on acoustics, speech, and signal processing (ICASSP 2003), Hong Kong, China, April 2003
7. Akkaya K, Younis M (2005) A survey on routing protocols for wireless sensor networks. Ad Hoc Netw 3(3):325–349
8. Abbasi AA, Younis M (2007) A survey on clustering algorithms for wireless sensor networks. Comput Commun 30(14):2826–2841
9. Tyagi S, Kumar N (2013) A systematic review on clustering and routing techniques based upon LEACH protocol for wireless sensor networks. J Netw Comput Appl 36(2):623–645
10. Pantazis N, Nikolidakis SA, Vergados DD (2013) Energy-efficient routing protocols in wireless sensor networks: a survey. Commun Surv Tutorials IEEE 15(2):551–591
11. Al-Karaki JN, Kamal AE (2004) Routing techniques in wireless sensor networks: a survey. Wirel Commun IEEE 11(6):6–28
12. Anastasi G et al (2009) Energy conservation in wireless sensor networks: a survey. Ad Hoc Netw 7(3):537–568
13. Radi M et al (2012) Multipath routing in wireless sensor networks: survey and research challenges. Sensors 12(1):650–685

14. Sendra S et al (2011) Power saving and energy optimization techniques for wireless sensor networks. J Commun 6(6):439–459
15. Min R, Bhardwaj R et al (2000) An Architecture for a power aware distributed micro sensor node. In: The proceedings of the IEEE workshop on signal processing systems (SIPS'00), October 2000
16. Heinzelman WR, Sinha A, Wang A et al (2000) Energy-scalable algorithms and protocols for wireless sensor networks. In: The proceedings of the international conference on acoustics, speech, and signal processing (ICASSP '00), Istanbul, Turkey, June 2000
17. Woo A, Culler D (2001) A transmission control scheme for media access in sensor networks. In: The proceedings of the 7th annual ACM/IEEE international conference on mobile computing and networking (Mobicom '01), Rome, Italy, July 2001
18. Shih E, Cho SH, Iceks N et al (2001) Physical layer driven protocol and algorithm design for energy-efficient wireless sensor networks. In: The proceedings of the 7th annual ACM/IEEE international conference on mobile computing and networking (Mobicom '01), Rome, Italy, July 2001
19. Ye W, Heidemann J, Estrin D (2002) An energy-efficient MAC protocol for wireless sensor networks. In: The proceedings of IEEE infocom 2002, New York, NY, June 2002
20. Bara'a AA, Khalil EA (2012) A new evolutionary based routing protocol for clustered heterogeneous wireless sensor networks. Appl Soft Comput 12(7) (July 2012):1950–1957. doi:10.1016/j.asoc.2011.04.007; http://dx.doi.org/10.1016/j.asoc.2011.04.007
21. Zheng J, Jamalipour A (2009) Wireless sensor networks: a networking perspective. Wiley, New York
22. Dulman S, Nieberg T, Wu J, Havinga P (2003) Trade—off between traffic overhead and reliability in multipath routing for wireless sensor networks. In: Proceedings IEEE WCNC' 03, New Orleans, LA, Mar 2003, pp 1918–1922
23. Girão PS, Enache GA (2007) Wireless sensor networks: state of the art and future trends. Instituto de Telecomunicações/Instituto Superior Técnico
24. Yi W, Chen Y (2012) Cross-layer design in wireless sensor networks. In: Rashvand HF, Kavian YS (eds) Using cross-layer techniques for communication systems. IGI Books, pp 527–565. doi:10.4018/978-1-4666-0960-0.ch021

Index

A
Applications, 2, 3, 5, 6, 9, 11
Architecture, 5, 8

C
Challenges, 2
Classification, 17
Coherent, 30, 39, 40, 43
Comparative analysis, 61
Comparison, 21, 54

D
Design issues, 2, 5, 9

E
Energy efficiency, 17, 21

F
Flat, 46–49, 54
Fundamentals, 59
Future trends, 27, 64

H
Hierarchical, 46, 47, 51, 52, 54

L
Location based, 46, 49, 51, 53, 54

M
Multi-hop routing, 29, 40, 43, 63

N
Network operations, 29–31, 40, 43
Network structure, 46, 55

Q
Quality of service (QoS), 40, 43
Query, 30, 33–35, 43

R
Reliability, 29–31, 33, 37, 40, 43
Routing, 2, 5–7, 9–11, 45, 47–50, 53, 54, 57

W
WSN, 1–3, 5, 6, 8, 9, 11, 15, 17, 20–22, 27

© The Author(s) 2016   69
S. Rani and S.H. Ahmed, *Multi-hop Routing in Wireless Sensor Networks*,
SpringerBriefs in Electrical and Computer Engineering,
DOI 10.1007/978-981-287-730-7

Printed in the United States
By Bookmasters